T0332647

Deep Learning and Linguistic Representation

Chapman & Hall/CRC Machine Learning & Pattern Recognition

Introduction to Machine Learning with Applications in Information Security
Mark Stamp

A First Course in Machine Learning
Simon Rogers, Mark Girolami

Statistical Reinforcement Learning: Modern Machine Learning Approaches
Masashi Sugiyama

Sparse Modeling: Theory, Algorithms, and Applications
Irina Rish, Genady Grabarnik

Computational Trust Models and Machine Learning
Xin Liu, Anwitaman Datta, Ee-Peng Lim

Regularization, Optimization, Kernels, and Support Vector Machines
Johan A.K. Suykens, Marco Signoretto, Andreas Argyriou

Machine Learning: An Algorithmic Perspective, Second Edition
Stephen Marsland

Bayesian Programming
Pierre Bessiere, Emmanuel Mazer, Juan Manuel Ahuactzin, Kamel Mekhnacha

Multilinear Subspace Learning: Dimensionality Reduction of Multidimensional Data
Haiping Lu, Konstantinos N. Plataniotis, Anastasios Venetsanopoulos

Data Science and Machine Learning: Mathematical and Statistical Methods
Dirk P. Kroese, Zdravko Botev, Thomas Taimre, Radislav Vaisman

Deep Learning and Linguistic Representation
Shalom Lappin

For more information on this series please visit: https://www.routledge.com/Chapman--Hall-CRC-Machine-Learning--Pattern-Recognition/book-series/CRCMACLEAPAT

Deep Learning and Linguistic Representation

Shalom Lappin

CRC Press is an imprint of the
Taylor & Francis Group, an **informa** business
A CHAPMAN & HALL BOOK

First edition published 2021
by CRC Press
6000 Broken Sound Parkway NW, Suite 300, Boca Raton, FL 33487-2742

and by CRC Press
2 Park Square, Milton Park, Abingdon, Oxon, OX14 4RN

Library of Congress Cataloging-in-Publication Data

Names: Lappin, Shalom, author.
Title: Deep learning and linguistic representation / Shalom Lappin.
Description: Boca Raton : CRC Press, 2021. | Includes bibliographical references and index.
Identifiers: LCCN 2020050622 | ISBN 9780367649470 (hardback) | ISBN 9780367648749 (paperback) | ISBN 9781003127086 (ebook)
Subjects: LCSH: Computational linguistics. | Natural language processing (Computer science) | Machine learning.
Classification: LCC P98 .L37 2021 | DDC 410.285--dc23
LC record available at https://lccn.loc.gov/2020050622

ISBN: 978-0-367-64947-0 (hbk)
ISBN: 978-0-367-64874-9 (pbk)
ISBN: 978-1-003-12708-6 (ebk)

Typeset in Latin Modern font
by KnowledgeWorks Global Ltd.

לזכר אמי עדה לפין, שלימדה אותי מהי אהבת שפה

Contents

Preface

Over the past 15 years deep learning has produced a revolution in artificial intelligence. In natural language processing it has created robust, large coverage systems that achieve impressive results across a wide range of applications, where these were resistant to more traditional machine learning methods, and to symbolic approaches. Deep neural networks (DNNs) have become dominant throughout many domains of AI in general, and in NLP in particular, by virtue of their success as engineering techniques.

Recently, a growing number of computational linguists and cognitive scientists have been asking what deep learning might teach us about the nature of human linguistic knowledge. Unlike the early connectionists of the 1980s, these researchers have generally avoided making claims about analogies between deep neural networks and the operations of the brain. Instead, they have considered the implications of these models for the cognitive foundations of natural language, in nuanced and indirect ways. In particular, they are interested in the types of syntactic structure that DNNs identify, and the semantic relations that they can recognise. They are concerned with the manner in which DNNs represent this information, and the training procedures through which they obtain it.

This line of research suggests that it is worth exploring points of similarity and divergence between the ways in which DNNs and humans encode linguistic information. The extent to which DNNs approach (and, in some cases, surpass) human performance on linguistically interesting NLP tasks, through efficient learning, gives some indication of the capacity of largely domain general computational learning devices for language learning. An obvious question is whether humans could, in principle, acquire this knowledge through similar sorts of learning processes.

This book draws together work on deep learning applied to natural language processing that I have done, together with colleagues, over the past eight years. It focusses on the question of what current methods of machine learning can contribute to our understanding of the way in which humans acquire and represent knowledge of the syntactic and semantic properties of their language. The book developed out of two online courses that I gave recently. I presented the first as a series of talks for students and colleagues from the Centre for Linguistic Theory and Studies in Probability (CLASP) at the University of Gothenburg, in June 2020. The second was an invited course for the Brandeis Web Summer School in Logic, Language, and Information, in July 2020. I am grateful to the participants of both forums for stimulating discussion and helpful comments.

I am deeply indebted to the colleagues and students with whom I did the joint experimental work summarised here. I use it as the basis for addressing the broader cognitive issues that constitute the book's focus. As will become clear from the co-authored publications that I cite throughout the following chapters, they have played a central role in the development of my ideas on these issues. I am enormously grateful to them for guiding me through much of the implementation of the work that we have done together. I wish to express my appreciation to Carlos Armendariz, Jean-Philippe Bernardy, Yuri Bizzoni, Alex Clark, Adam Ek, Gianluca Giorgolo, Jey Han Lau, and Matt Purver for their very major contributions to our joint work.

The students and research staff at CLASP have provided a wonderfully supportive research environment. Their scientific activity and, above all, their friendship, have played a significant role in facilitating the work presented here. Stergios Chatzikyriakidis and Bill Noble have been an important source of feedback for the development of some of the ideas presented here. I am grateful to my colleagues in the Cognitive Science Group in the School of Electronic Engineering and Computer Science at Queen Mary University of London for helpful discussion of some of the questions that I take up in this book. Stephen Clark and Pat Healey have provided helpful advice on different aspects of my recent research. I would also like to thank Devdatt Dubhashi, head of the Machine Learning Group at Chalmers University of Science in Gothenburg, for introducing me to many of the technical aspects of current work in deep learning, and for lively discussion of its relevance to NLP. I am particularly grateful to Stephen Clark for detailed comments on an earlier draft of this monograph. He caught many mistakes, and suggested valuable improvements. Needless to say, I bear sole responsibility for any errors in this book. My work on the monograph is supported by grant 2014-39 from the Swedish Research Council, which funds CLASP.

Elliott Morsia, my editor at Taylor and Francis, has provided superb help and support. Talitha Duncan-Todd, my production person, and

Shashi Kumar, the Latex support person, have given me much needed assistance throughout the production process.

My family, particularly my children and grandchildren, are the source of joy and wider sense of purpose needed to complete this, and many other projects. While they share the consensus that I am a hopeless nerd, they assure me that the scientific issues that I discuss here are worthwhile. They frequently ask thoughtful questions that help to advance my thinking on these issues. They remain permanently surprised that someone so obviously out of it could work in such a cool field. In addition to having them, this is indeed one of the many blessings I enjoy. Above all, my wife Elena is a constant source of love and encouragement. Without her none of this would be possible.

The book was written in the shadow of the covid 19 pandemic. This terrible event has brought at least three phenomena clearly into view. First, it has underlined the imperative of taking the results of scientific research seriously, and applying them to public policy decisions. Leaders who dismiss well motivated medical advice, and respond to the crisis through denial and propaganda, are inflicting needless suffering on their people. By contrast, governments that allow themselves to be guided by well supported scientific work have been able to mitigate the damage that the crisis is causing.

Second, the crisis has provided a case study in the damage that large scale campaigns of disinformation, and defamation, can cause to the health and the well-being of large numbers of people. Unfortunately, digital technology, some of it involving NLP applications, has provided the primary devices through which these campaigns are conducted. Computer scientists working on these technologies have a responsibility to address the misuse of their work for socially destructive purposes. In many cases, this same technology can be applied to filter disinformation and hate propaganda. It is also necessary to insist that the agencies for which we do this work be held accountable for the way in which they use it.

Third, the pandemic has laid bare the devastating effects of extreme economic and social inequality, with the poor and ethnically excluded bearing the brunt of its effects. Nowhere has this inequality been more apparent than in the digital technology industry. The enterprises of this industry sustain much of the innovative work being done in deep learning. They also instantiate the sharp disparities of wealth, class, and opportunity that the pandemic has forced into glaring relief. The engineering and scientific advances that machine learning is generating hold out the promise of major social and environmental benefit. In order for this promise to be realised, it is necessary to address the acute deficit in democratic accountability, and in equitable economic arrangements that the digital technology industry has helped to create.

Scientists working in this domain can no longer afford to treat these problems as irrelevant to their research. The survival and stability of the societies that sustain this research depend on finding reasonable solutions to them.

NLP has blossomed into a wonderfully vigorous field of research. Deep learning is still in its infancy, and it is likely that the architectures of its systems will change radically in the near future. By using it to achieve perspective on human cognition, we stand to gain important insight into linguistic knowledge. In pursuing this work it is essential that we pay close attention to the social consequences of our scientific research.

Shalom Lappin
London
October, 2020

CHAPTER 1

Introduction: Deep Learning in Natural Language Processing

1.1 OUTLINE OF THE BOOK

In this chapter I will briefly introduce some of the main formal and architectural elements of deep learning systems.[1] I will provide an overview of the major types of DNN used in NLP. We will start with simple feed forward networks and move on to different types of Recurrent Neural Networks (RNNs), specifically, simple RNNs and Long Short-Term Memory RNNs. We will next look at Convolutional Neural Networks (CNNs), and then conclude with Transformers. For the latter type of network we will consider GPT-2, GPT-3, and BERT. I conclude the chapter with a composite DNN that Bizzoni and Lappin (2017) construct for paraphrase assessment, in order to illustrate how these models are used in NLP applications.

Chapter 2 is devoted to recent work on training DNNs to learn syntactic structure for a variety of tasks. The first application that I look at is predicting subject-verb agreement across sequences of possible NP controllers (Bernardy & Lappin, 2017; Gulordava, Bojanowski, Grave, Linzen, & Baroni, 2018; Linzen, Dupoux, & Goldberg, 2016). It is necessary to learn hierarchical syntactic structure to succeed at this task, as linear order does not, in general, determine subject-verb agreement. An

[1]For an excellent detailed guide to the mathematical and formal concepts of deep learning consult Goodfellow, Bengio, and Courville (2016).

important issue here is whether unsupervised neural language models (LMs) can equal or surpass the performance of supervised LSTM models. I then look at work comparing Tree RNNs, which encode syntactic tree structure, with sequential (non-tree) LSTMs, across several other applications.

In Chapter 3, I present work by Lau, Clark, and Lappin (2017) on using a variety of machine learning methods, including RNNs, to predict mean human sentence acceptability judgements. They give experimental evidence that human acceptability judgements are individually, as well as aggregately, gradient, and they test several machine learning models on crowd-sourced annotated corpora. These include naturally occurring text from the British National Corpus (BNC) and Wikipedia, which is subjected to round-trip machine translation through another language, and back to English, to introduce infelicities into the test corpus. Lau, Clark, and Lappin (2017) extend these experiments to other languages. They also test their models on a set of linguists' examples, which they annotate through crowd sourcing. I conclude this chapter with a discussion of Ek, Bernardy, and Lappin (2019), which reports the results of an experiment in which LSTMs are trained on Wikipedia corpora enriched with a variety of syntactic and semantic markers, and then tested for predicting mean human acceptability judgements, on Lau, Clark, and Lappin (2017)'s annotated BNC test set. The sentence acceptability task is linguistically interesting because it measures the capacity of machine learning language models to predict the sorts of judgements that have been widely used to motivate linguistic theories. The accuracy of a language model in this task indicates the extent to which it can acquire the sort of knowledge that speakers use in classifying sentences as more or less grammatical and well-formed in other respects.

Chapter 4 looks at recent work by Bernardy, Lappin, and Lau (2018) and Lau, Armendariz, Lappin, Purver, and Shu (2020) on extending the sentence acceptability task to predicting mean human judgements of sentences presented in different sorts of document contexts. The crowd-source experiments reported in these papers reveal an unexpected compression effect, in which speakers assessing sentences in both real and random contexts raise acceptability scores, relative to out of context ratings, at the lower end of the scale, but lower them at the high end. Lau et al. (2020) control for a variety of confounds in order to identify the factors that seem to produce the compression effect. This chapter also presents the results of Lau's new total least squares regression work, which confirms that this effect is a genuine property of the data, rather than regression to the mean.

Lau et al. (2020) expand the set of neural language models to include unidirectional and bidirectional transformers. They find that bidirectional, but not unidirectional, transformers approach a plausible estimated upper bound on individual human prediction of sentence acceptability, across context types. This result raises interesting questions concerning the role of directionality in human sentence processing.

In Chapter 5 I discuss whether DNNs, particularly those described in previous chapters, offer cognitively plausible models of linguistic representation and language acquisition. I suggest that if linguistic theories provide accurate explanations of linguistic knowledge, then NLP systems that incorporate their insights should perform better than those that do not, and I explore whether these theories, specifically those of formal syntax, have, in fact, made significant contributions to solving NLP tasks. Answering this question involves looking at more recent DNNs enriched with syntactic structure. I also compare DNNs with grammars, as models of linguistic knowledge. I respond to criticisms that Sprouse, Yankama, Indurkhya, Fong, and Berwick (2018) raise against Lau, Clark, and Lappin (2017)'s work on neural language models for the sentence acceptability task to support the view that syntactic knowledge is probabilistic rather than binary in nature. Finally, I consider three well-known cases from the history of linguistics and cognitive science in which theorists reject an entire class of models as unsuitable for encoding human linguistic knowledge, on the basis of the limitations of a particular member of the class. The success of more sophisticated models in the class has subsequently shown these inferences to be unsound. They represent influential cases of over reach, in which convincing criticism of a fairly simple computational model is used to dismiss all models of a given type, without considering straightforward improvements that avoid the limitations of the simpler system.

I conclude Chapter 5 with a discussion of the application of deep learning to distributional semantics. I first briefly consider the type theoretic model that Coecke, Sadrzadeh, and Clark (2010) and Grefenstette, Sadrzadeh, Clark, Coecke, and Pulman (2011) develop to construct compositional interpretations for phrases and sentences from distributional vectors, on the basis of the syntactic structure specified by a pregroup grammar. This view poses a number of conceptual and empirical problems. I then suggest an alternative approach on which semantic interpretation in a deep learning context is an instance of sequence to sequence (seq2seq) machine translation. This involves mapping sentence vectors into multimodal vectors that represent non-linguistic events and situations.

Chapter 6 presents the main conclusions of the book. It briefly takes up some of the unresolved issues, and the questions raised by the research discussed here. I consider how to pursue these in future work.

1.2 FROM ENGINEERING TO COGNITIVE SCIENCE

Over the past ten years the emergence of powerful deep learning (DL) models has produced significant advances across a wide range of AI tasks and domains. These include, among others, image classification, face recognition, medical diagnostics, game playing, and autonomous robots. DL has been particularly influential in NLP, where it has yielded substantial progress in applications like machine translation, speech recognition, question-answering, dialogue management, paraphrase identification and natural language inference (NLI). In these areas of research, it has displaced other machine learning methods to become the dominant approach.

The success of DL as an engineering method raises important cognitive issues. DNNs constitute domain general learning devices, which apply the same basic approach to learning, data processing, and representation to all types of input data. If they are able to approximate or surpass human performance in a task, what conclusions, if any, can we draw concerning the nature of human learning and representation for that task?

Lappin and Shieber (2007) and A. Clark and Lappin (2011) suggest that computational learning theory provides a guide to determining how much linguistic knowledge can be acquired through different types of machine learning models. They argue that relatively weak bias models can efficiently learn complex grammar classes suitable for natural language syntax. Their results do not entail that humans actually use these models for language acquisition. But they do indicate the classes of grammars that humans can, in principle, acquire through domain general methods of induction, from tractable quantities of data, in plausible amounts of time.

Early connectionists (Rumelhart, McClelland, & PDP Research Group, 1986) asserted that neural networks are modelled on the human brain. Few people working in DL today make this strong claim. The extent, if any, to which human learning and representation resemble those of a

DNN can only be determined by neuroscientific research. A weak view of DL takes it to show what sorts of knowledge can be acquired by domain general learning procedures. To the degree that domain-specific learning biases must be added to a DNN, either through architectural design or enrichment of training data with feature annotation, in order to succeed at an AI task, domain general learning alone is not sufficient to achieve knowledge of this task.

The distinction between strong and weak views of DNNs is tangentially related to the difference between strong and weak AI. On the strong view of AI, an objective of research in artificial intelligence is to construct computational agents that reproduce general human intelligence and are fully capable of human reasoning. The weak view of AI takes its objective to be the development of computational devices that achieve functional equivalence to human problem solving abilities.[2] But while related, these two pairs of notions are distinct, and it is important not to confuse them. The strong vs weak construal of DL turns not on the issue of replicating general intelligence, but on whether or not one regards DNNs as models of the brain.

In this book I am setting aside the controversy between strong vs weak AI, while adopting the weak view of DL. If a DNN is able to approach or surpass human performance on a linguistic task, then this shows how domain general learning mechanisms, possibly supplemented with additional domain bias factors, can, in principle, acquire this knowledge efficiently. I am not concentrating on grammar induction, but the more general issues of language learning and the nature of linguistic representation.

Many linguists and cognitive scientists express discomfort with DNNs as models of learning and representation, on the grounds that they are opaque in the way in which they produce their output. They are frequently described as black box devices that are not accessible to clear explanation, in contrast to more traditional machine learning models and symbolic, rule-based systems. Learning theorists also observe that older computational learning theories, like PAC (Valiant, 1984) and Bayesian models (Berger, 1985), prove general results on the class of learnable

[2]See Bringsjord and Govindarajulu (2020) for a recent discussion of strong and weak approaches to AI. Turing (1950) played a significant role in defining the terms of debate between these two views of AI by suggesting a test of general intelligence on which humans would be unable to distinguish an artificial from a human interlocutor in dialogue. For discussions of the Turing test, see Shieber (2007) and the papers in Shieber (2004).

objects. These results specify the rates and complexity of learning, in proportion to resources of time and data, for their respective frameworks.[3] In general, such results are not yet available for different classes of DNN.

It is certainly the case that we still have much to discover about how DNNs function and the formal limits of their capacities. However, it is not accurate to describe them as sealed apparatuses whose inner working is shrouded in mystery. We design their architecture, implement them and control their operation. A growing number of researchers are devising methods to illuminate the internal patterns through which DNNs encode, store, and process information of different kinds. The papers in Linzen, Chrupała, and Alishahi (2018) and Linzen, Chrupała, Belinkov, and Hupkes (2019) address the issue of explainability of DL in NLP, as part of an ongoing series of workshops devoted to this question.

At the other extreme one encounters hyperbole and unrealistic expectations that DL will soon yield intelligent agents able to engage in complex reasoning and inference, at a level that equals or surpasses human performance. This has given rise to fears of powerful robots that can become malevolent AI actors.[4] This view is not grounded in the reality of current AI research. DNNs have not achieved impressive results in domain general reasoning tasks. A breakthrough in this area does not seem imminent with current systems, nor is it likely in the foreseeable future. The relative success of DNNs with natural language inference test suites relies on pattern matching and analogy, achieved through training on large data sets of the same type as the sets on which they are tested. Their performance is easily disrupted by adversarial testing with substitutions of alternative sentences in these test sets. Talman and Chatzikyriakidis (2019) provide interesting experimental results of this sort of adversarial testing.

DL is neither an ad hoc engineering tool that applies impenetrable processing systems to vast amounts of data to achieve results through inexplicable operations, nor a technology that is quickly approaching the construction of agents with powerful general intelligence. It is a class of remarkably effective learning models that have made impressive advances across a wide range of AI applications. By studying the abilities and the limitations of these models in handling linguistically interesting

[3]See Lappin and Shieber (2007), and A. Clark and Lappin (2011) for the application of PAC and statistical learning theory to grammar induction.

[4]See Dubhashi and Lappin (2017) for arguments that these concerns are seriously misplaced.

NLP tasks, we stand to gain useful insights into possible ways of encoding and representing natural language as part of the language learning process. We will also deepen our understanding of the relative contributions of domain general induction procedures on one hand, and language specific learning biases on other, to the success and efficiency of this process.

1.3 ELEMENTS OF DEEP LEARNING

DNNs learn an approximation of a function $f(x) = y$ which maps input data x to an output value y. These values generally involve classifying an object relative to a set of possible categories or determining the conditional probability of an event, given a sequence of preceding occurrences. Deep Feed Forward Networks consist of

(i) an input layer where data are entered,

(ii) one or more hidden layers, in which units (neurons) compute the weights for components of the data, and

(iii) an output layer that generates a value for the function.

DNNs can, in principle, approximate any function and, in particular, non-linear functions. Sigmoid functions are commonly used to determine the activation threshold for a neuron. The graph in Fig 1.1 represents a set of values that a sigmoid function specifies. The architecture of a feed forward network is shown in Fig 1.2.

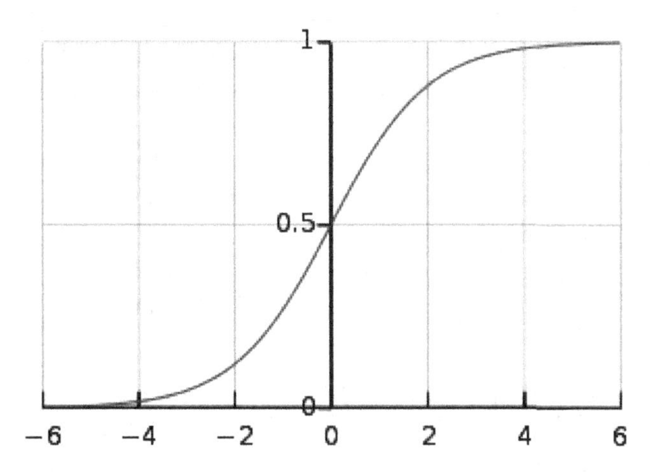

Figure 1.1 Sigmoid function.

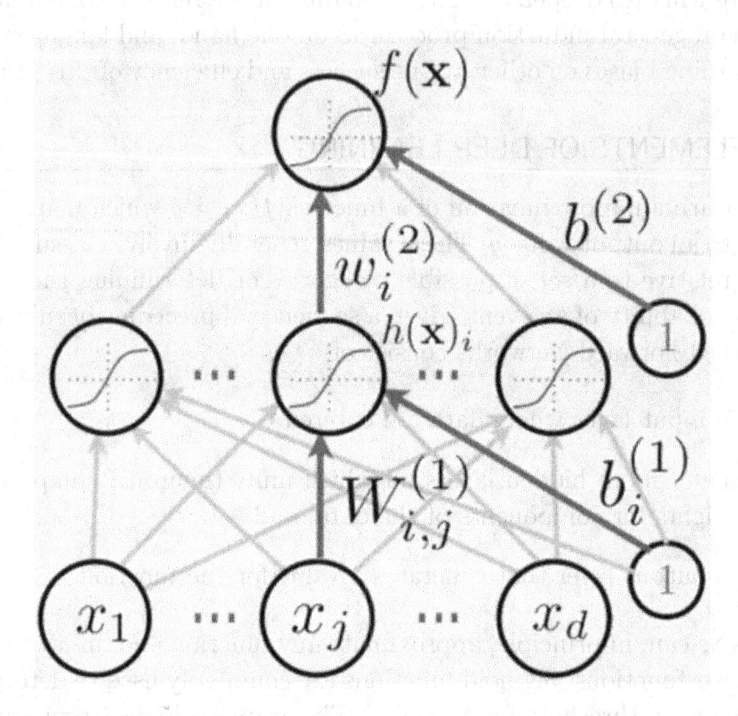

Figure 1.2 Feed Forward Network.
From Tushar Gupta, "Deep Learning: Feedforward Neural Network", *Towards Data Science*, January 5, 2017.

Training a DNN involves comparing its predicted function value to the ground truth of its training data. Its error rate is reduced in cycles (epochs) through back propagation. This process involves computing the gradient of a loss (error) function and proceeding down the slope, by specified increments, to an estimated optimal level, determined by stochastic gradient descent.

Cross Entropy is a function that measures the difference between two probability distributions P and Q through the formula:

$$H(P,Q) = -E_{x \sim P}\ log\ Q(x)$$

$H(P,Q)$ is the cross entropy of the distribution Q relative to the distribution P. $-E_{x \sim P}\ log\ Q(x)$ is the negative of the expected value, for x, given P, of the natural logarithm of $Q(x)$. Cross entropy is widely used as a loss function for gradient descent in training DNNs. At each epoch in the training process cross entropy is computed, and the values of the weights assigned to the hidden units are adjusted to reduce error along the slope identified by gradient descent. Training is concluded when the distance between the network's predicted distribution and that projected from the training data reach an estimated optimal minimum.

In many cases the hidden layers of a DNN will produce a set of non-normalised probability scores for the different states of a random variable corresponding to a category judgement, or the likelihood of an event. The softmax function maps the vector of these scores into a normalised probability distribution whose values sum to 1. The function is defined as follows:

$$softmax(z)_i = \frac{e^{z_i}}{\sum_j e^{z_j}}$$

This function applies the exponential function to each input value, and normalises it by dividing it with the sum of the exponentials for all the inputs, to insure that the output values sum to 1. The softmax function is widely used in the output layer of a DNN to generate a probability distribution for a classifier, or for a probability model.

Words are represented in a DNN by vectors of real numbers. Each element of the vector expresses a distributional feature of the word. These features are the dimensions of the vectors, and they encode its co-occurrence patterns with other words in a training corpus. Word embeddings are generally compressed into low dimensional vectors (200–300 dimensions) that express similarity and proximity relations among the words in the vocabulary of a DNN model. These models frequently use large pre-trained word embeddings, like word2vec (Mikolov, Kombrink, Deoras, Burget, & Èernocký, 2011) and GloVe (Pennington, Socher, & Manning, 2014), compiled from millions of words of text.

In supervised learning a DNN is trained on data annotated with the features that it is learning to predict. For example, if the DNN is learning to identify the objects that appear in graphic images, then its training data may consist of large numbers of labelled images of the objects that it is intended to recognise in photographs. In unsupervised learning the training data are not labelled.[5] A generative neural language model may be trained on large quantities of raw text. It will generate the most likely word in a sequence, given the previous words, on the basis of the probability distribution over words, and sequences of words, that it estimates from the unlabelled training corpus.

1.4 TYPES OF DEEP NEURAL NETWORKS

Feed Forward Neural Networks take data encoded in vectors of fixed size as input, and they yield output vectors of fixed size. Recurrent Neural Networks (RNNs) (Elman, 1990) apply to sequences of input vectors, producing a string of output vectors. They retain information from previous processing phases in a sequence, and so they have a memory over the span of the input. RNNs are particularly well suited to processing natural language, whose units of sound and text are structured as ordered strings. Fig 1.3 shows the architecture of an RNN.

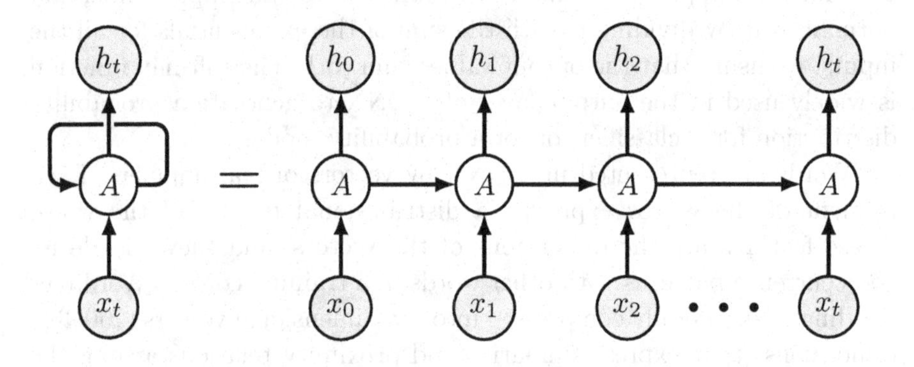

Figure 1.3 Recurrent Neural Network

Simple RNNs preserve information from previous states, but they do not effectively control this information. They have difficulties representing long-distance dependencies between elements of a sequence. A

[5]See A. Clark and Lappin (2010) for a detailed discussion of supervised and unsupervised learning in NLP.

Long Short-Term Memory network (Hochreiter & Schmidhuber, 1997) is a type of RNN whose units contain three types of information gates, composed of sigmoid and hyperbolic tangent (tanh) functions.[6]

(i) The forgetting gate determines which part of the information received from preceding units is discarded;

(ii) the input gate updates the retained information with the features of a new element of the input sequence; and

(iii) the output gate defines the vector which is passed to the next unit in the network.

Fig 1.4 displays the architecture of an LSTM.

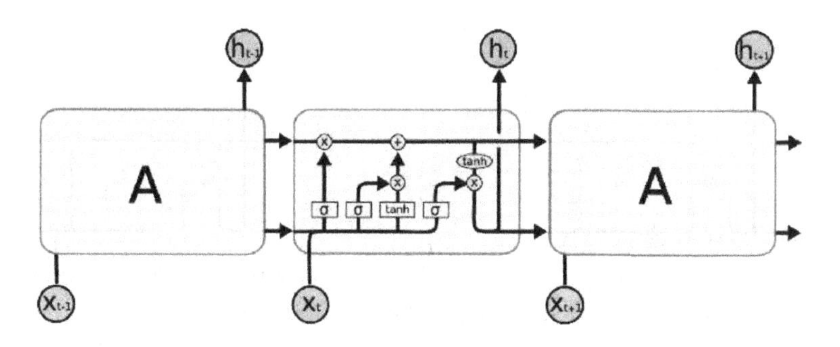

Figure 1.4 LSTM.
From Christopher Olah's blog *Understanding LSTM Networks*, August 27, 2015.

In a convolutional neural network (CNN, Lecun, Kavukcuoglu, & Farabet, 2010) input is fed to a convolutional layer, which extracts a

[6]Where a logistic sigmoid function is a sigmoid function that generates values from 0 to 1, which can be interpreted as probabilities, tanh is a rescaled logistic sigmoid function that returns outputs from −1 to 1. A logistic sigmoid function is defined by the equation

$$S(x) = \frac{e^x}{e^x + 1}.$$

A tanh function is defined by the equation

$$tanh\ z = \frac{e^{2z} - 1}{e^{2z} + 1}.$$

feature map from this data. A pooling layer compresses the map by reducing its dimensions, and rendering it invariant to small changes in input (noise filtering). Successive convolutional + pooling layers construct progressively higher level representations from the feature maps received from preceding levels of the network. The output feature map is passed to one or more fully interconnected layers, which transform the map into a feature vector. A softmax function maps this vector into a probability distribution over the states of a category variable. Fig 1.5 illustrates the structure of a CNN.[7]

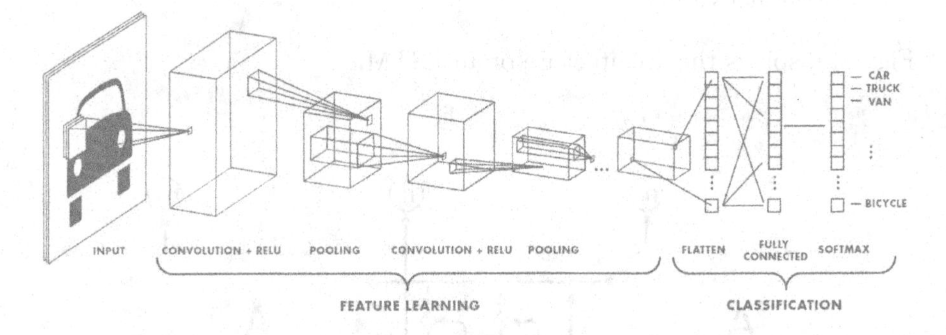

Figure 1.5 CNN.
From Sumit Saha "A Comprehensive Guide to Convolutional Neural Networks—the ELI5 Way", *Towards Data Science*, December 15, 2018.

Attention was developed to solve a problem in seq2seq neural machine translation, which uses an encoder-decoder architecture. In earlier versions of this architecture an RNN (or LSTM) encoded an input sequence as a single context vector, which a decoder RNN (LSTM) mapped to a target language sequence. Information from the previous hidden states of the encoder is lost, and all the words in the encoder's output vector are given equal weight when it is passed to the decoder.

Bahdanau, Cho, and Bengio (2015) introduce an attention layer that computes relative weights for each of the words in the input sequence, and these are combined with the context vector. The attention mechanism significantly improves the accuracy of seq2seq word alignment. It learns the relative importance of elements in the input in determining

[7]This diagram of a CNN appears in several publications, and on a number of websites. I have not managed to locate the original source for it.

correspondences to elements in the output sequence. Self-attention identifies relations among the elements of the same sequence, which enhances the capacity of the DNN to recognise long-distance dependency patterns in that sequence. Fig 1.6 displays an attention level interposed between an encoder (in this case, a bidirectional RNN) and a decoder (a unidrectional RNN). This component assigns different weights to the elements of the context vector, which the encoder produces as the input to the decoder.

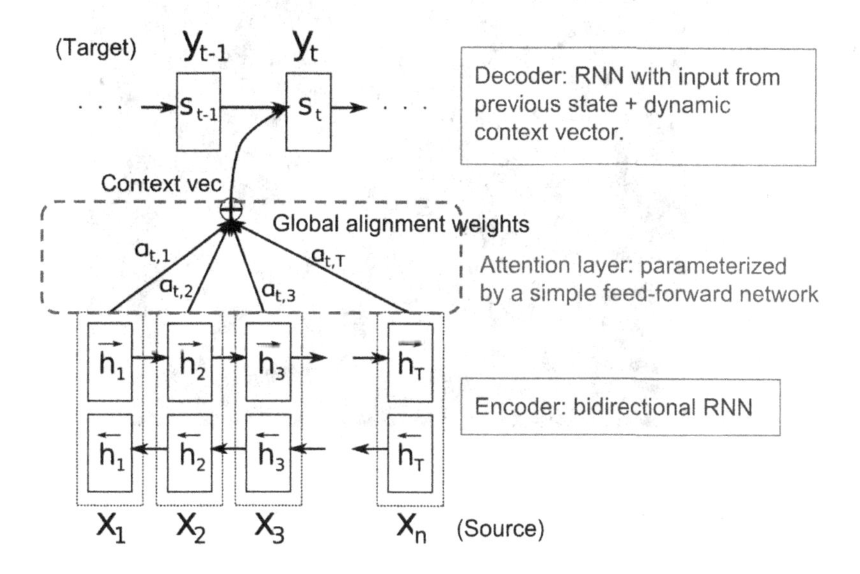

Figure 1.6 Encoder-Decoder System with Attention.
From Bahdanau et al. (2015).

Fig 1.7 is an attention alignment map for an English-French machine translation system. The number of squares and the degree of shading for word pairs, along the diagonal line of the map, indicate the relative attention allocated to these pairs.

Transformers (Vaswani et al., 2017) dispense with recurrent networks and convolution. Instead they construct both encoders and decoders out of stacks of layers that consist of multi-head attention units which provide input to a feed forward network. These layers process input sequences simultaneously, in parallel, independently of sequential order.

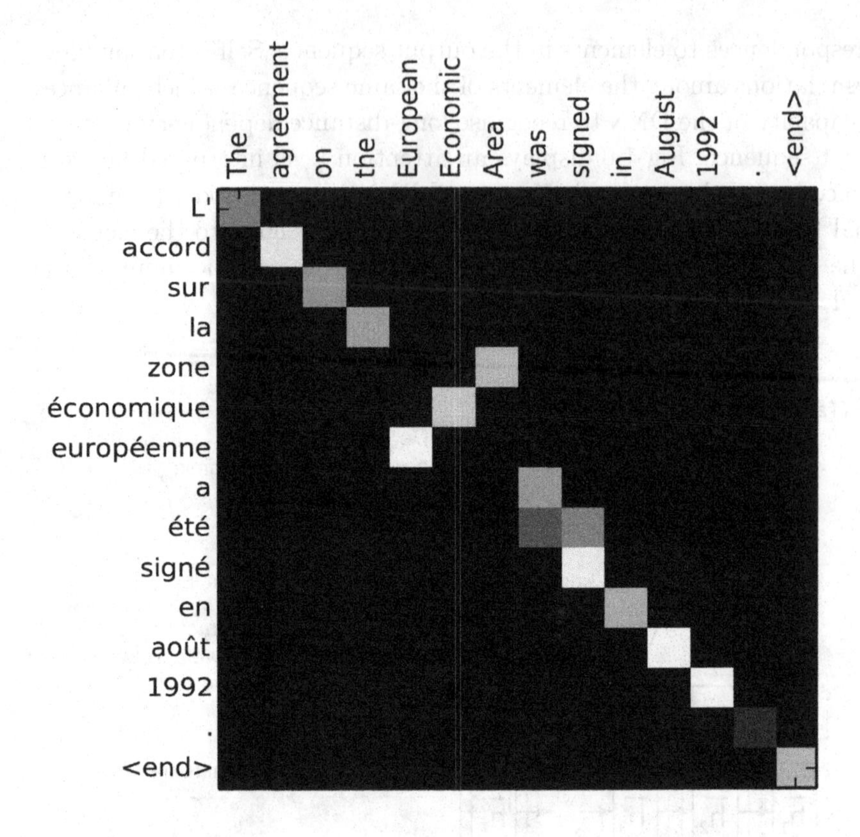

Figure 1.7 Attention Alignment for Machine Translation.
From Bahdanau et al. (2015).

However, the relative positions of the elements of a sequence are represented as additional information, which a transformer can exploit. The attention-driven design of transformers has allowed them to achieve significant improvements over LSTMs and CNNs, across a wide range of tasks. Fig 1.8 shows the architecture of a multi-head feed forward transformer.

Transformers are pre-trained on large amounts of text for extensive lexical embeddings. Many like OpenAI GPT (Radford, Narasimhan, Salimans, & Sutskever, 2018) have unidirectional architecture. GPT-2 (Solaiman et al., 2019) is a large transformer-based language model that OpenAI released in 2019. It is pre-trained on billions of words of text, with 1.5 billion parameters, where these support large-scale embeddings.

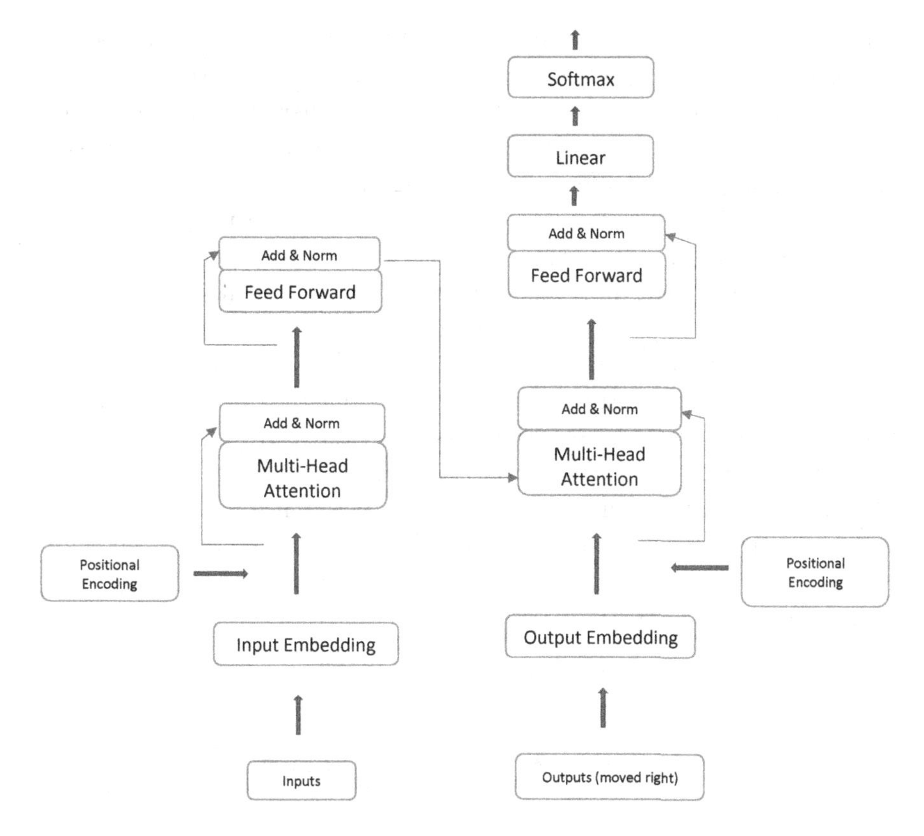

Figure 1.8 Architecture of a Transformer.

It is unidirectional in that the likelihood of a word in a sentence is conditioned by the likelihood of the words that precede it in the sequence. The probability of the sentence is the product of these probabilities, computed with the following equation.

$$P(s) = \prod_{i=0}^{|s|} P(w_i|w_{<i})$$

In 2020 OpenAI released GPT-3 (Brown et al., 2020). It retains GPT-2's architecture, but it is greatly increased in size, pre-trained for 175 billion parameters. It has been tested on a variety of NLP applications for learning, with limited exposure to examples of the solved task. Brown et al. (2020) report that GPT-3 shows promising results for zero-, one- and few-shot learning. In the latter case the model is exposed to 10–100 training examples. This is an important result because it shows that a large-scale pre-trained transformer can learn to perform some NLP tasks with very limited training. However, smaller transformer models that are fine-tuned on task-specific data still outperform GPT-3 on most of these tasks, even with few-shot learning. The notable exception is the generation of news reports. The rate at which human judges succeeded in identifying GPT-3 generated text as artificially produced was only slightly above chance. It is also important to keep in mind that GPT-3 achieves gains in fast learning for new tasks, at the cost of massive training on large corpora.

BERT is a bidirectional transformer trained to predict a masked token from both its left and right contexts (effectively it predicts the word in a blank between two contexts). It also uses the same generic parameters from its training for each task to which it is applied, and it is then fine-tuned for a particular task. BERT's architecture is shown in Fig 1.9, and its training regimen is indicated in Fig 1.10 (Devlin, Chang, Lee, & Toutanova, 2019).

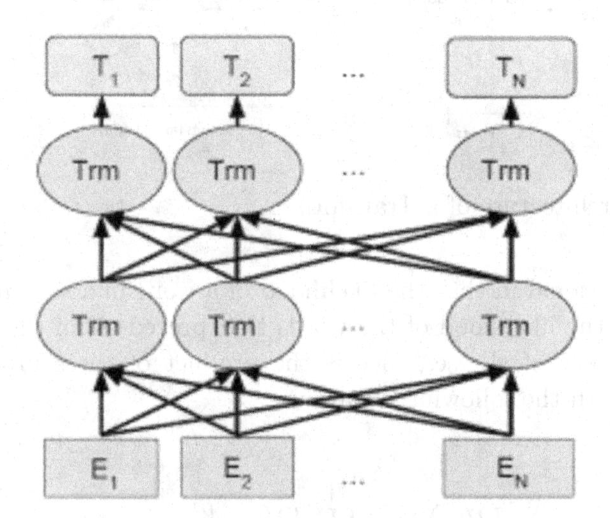

Figure 1.9 BERT.
From Devlin et al. (2019).

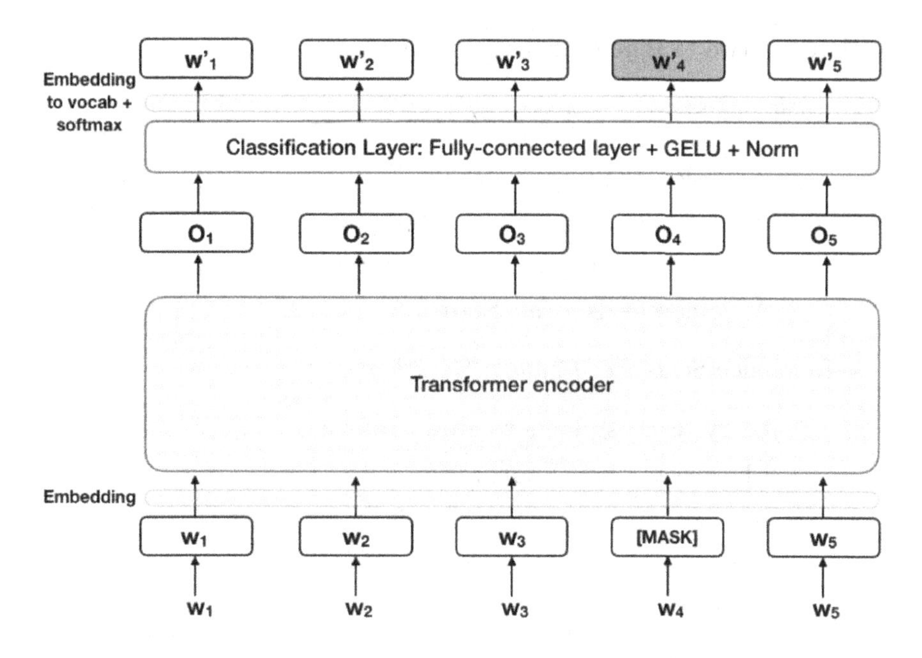

Figure 1.10 Training BERT.
From Rani Horev (2018), "BERT–State of the Art Language Model for NLP", *Lyrn.AI*, November 7, 2018.

1.5 AN EXAMPLE APPLICATION

Let's illustrate the use of DNNs in an NLP task, with some recent work on semantic paraphrase. Bizzoni and Lappin (2017) construct a composite neural network to classify sets of sentences for paraphrase proximity. They develop a corpus of 250 sets of five sentences, where each set contains a reference sentence and four paraphrase candidates. They rate each of the four candidates on a five-point scale for paraphrase proximity to the reference sentence. Every group of five sentences illustrates (possibly different) graduated degrees of the paraphrase relation within the reference sentence. Their rating labels correspond to the following categories; (1) two sentences are completely unrelated; (2) two sentences are semantically related, but they are not paraphrases; (3) two sentences are weak paraphrases; (4) two sentences are strong paraphrases; (5) two sentences are (type) identical.

Here are two examples of these ranking labels.

- Reference Sentence: *A woman feeds a cat*
    - A woman kicks a cat. **Score: 2**
    - A person feeds an animal. **Score: 3**
    - A woman is feeding a cat. **Score: 4**
    - A woman feeds a cat. **Score: 5**
- Reference Sentence: *I have a black hat*
    - Larry teaches plants to grow. **Score: 1**
    - I have a red hat. **Score: 2**
    - My hat is night black; pitch black. **Score: 3**
    - My hat's color is black. **Score: 4**

This annotation scheme sustains graded semantic similarity assessment, while also allowing for binary classification of a pair of sentences, scored independently of the other pairs in the reference set. The scores of two paraphrase candidates represent relative proximity to the reference sentence.

Bizzoni and Lappin (2017) train their classifier DNN for both binary and gradient classification of pairs of sentences for paraphrase. They train it on 761 pairs of sentences from the corpus, and they test it on 239 pairs.

The paraphrase classifier consists of three main components:

(i) two encoders, one for each of the sentences in a reference sentence-candidate pair, that consist of a CNN, a max pooling layer, and an LSTM,

(ii) a merge layer that concatenates the sentence vectors which the encoders produce into a single vector, and

(iii) several dense, fully connected layers that apply sigmoid functions to generate a softmax distribution for the paraphrase classification relation between the two input sentences.

The DNN uses the pre-trained lexical embeddings of Word2Vec (Mikolov et al., 2013). The CNN of the encoder identifies relevant features of an input sentence for the classification task. The max pooling

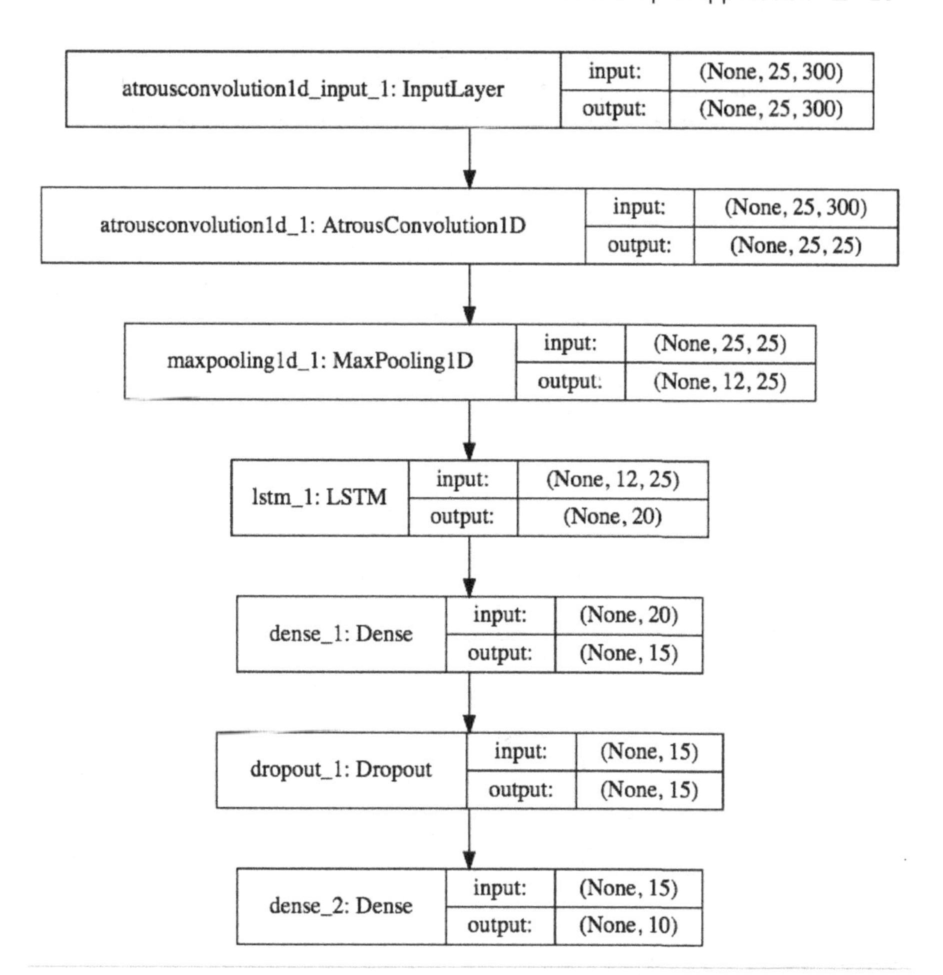

Figure 1.11 Paraphrase Encoder.
From Bizzoni and Lappin (2017).

layer reduces the dimensions of the vector that the CNN generates. The LSTM uses the sequential structure of the sentence vector to highlight features needed for the task, and to further reduce the dimensionality of the input vector. The LSTM produces a vector that is passed to two fully connected layers, the first one with a 0.5 dropout rate. Output from half the neurons, randomly selected, of this layer is discarded in training, to avoid overfitting. The structure of the paraphrase encoder is shown in Fig 1.11, and the composite classifier DNN is displayed in Fig 1.12.

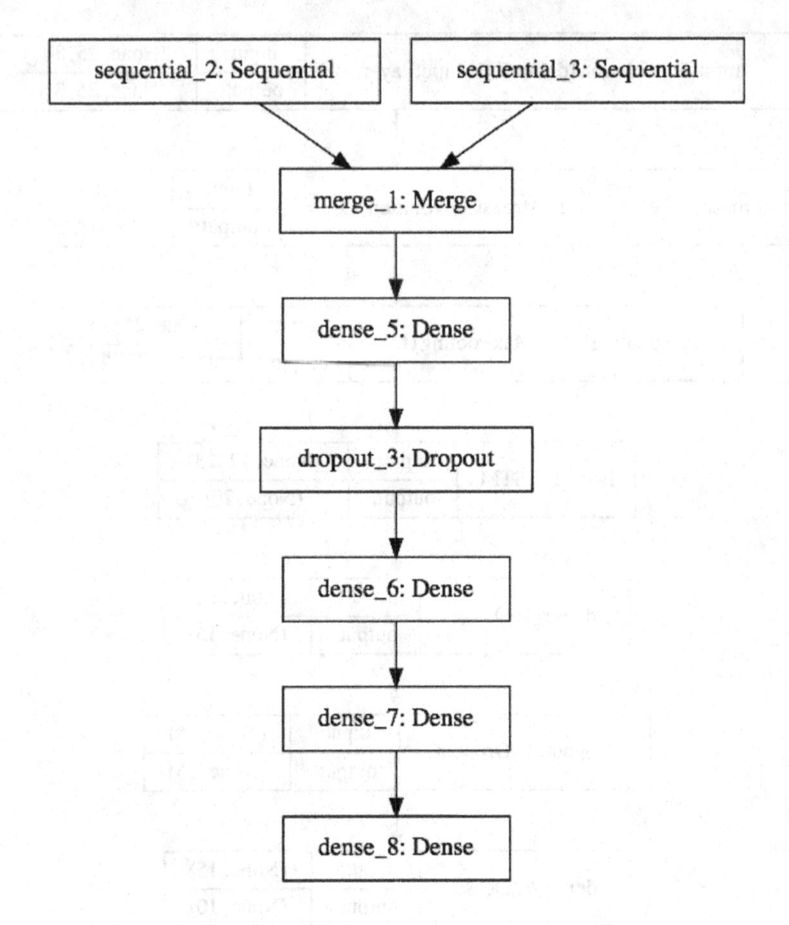

Figure 1.12 Paraphrase Classifier.
From Bizzoni and Lappin (2017).

Bizzoni and Lappin (2017) assess the accuracy of both binary and gradient classification on the basis of their annotation of the test set sentence pairs on a five-point scale. The binary classifier takes a softmax prediction of a score above a threshold of 2 as an instance of paraphrase. The gradient classifier predicts a paraphrase score from the scale through the softmax probability distribution over this relation. They use the Pearson coefficient to evaluate the correlation between the classifier's scores and the ground truth annotations.[8]

[8]The Pearson correlation coefficient (also known as Pearson's r) is a statistical measure of the linear correlation between two random variables X and Y. A positive value indicates the degree of positive correlation and a negative value, the extent of

TABLE 1.1 Binary Accuracy and Gradient Correlation for the Paraphrase Classifier

k	Accuracy
1	70.10
2	67.01
3	79.38
4	73.20
5	67.01
6	72.92
7	66.67
8	75.79
9	64.21
10	73.68
Average	71

k	Pearson
1	0.51
2	0.63
3	0.59
4	0.62
5	0.61
6	0.72
7	0.59
8	0.67
9	0.54
10	0.67
Average	0.61

They apply ten-fold cross-validation to test the robustness of accuracy and correlation. This involves successively partitioning the corpus into training and test components over ten different splits, in order to insure the robustness of test results. The accuracy and correlation scores for the paraphrase classifier are given in Table 1.1. Given the small size of the corpus, these are encouraging results for the classifier.

1.6 SUMMARY AND CONCLUSIONS

In this chapter we have briefly looked at the main types of DNN that are currently driving deep learning applications. We started with generic feed forward networks, and considered how back propagation, with gradient descent as an error reduction procedure, is used to train DNNs

a negative correlation. It is specified by the formula

$$\rho(X, Y) = \frac{cov(X, Y)}{\sigma_X \sigma_Y},$$

where ρ is the Pearson correlation between X and Y, cov is their covariance, and σ is their respective standard deviations. An alternative measure, the Spearman correlation, assesses the correlation between the ranked values of X and Y. These two statistical metrics may yield distinct correlation values for the same pairs of random values. However, in all of my joint work on DNNs predicting mean human judgements which I discuss here, we have found that Pearson and Spearman correlations converge on approximately the same values. Therefore, we report only Pearson correlations.

against a ground truth standard. We have seen how RNNs, enriched with long short-term memory through filtering and update functions, recognise patterns in sequences. CNNs use convolution and max pooling to progressively reduce the dimensions of feature vectors, and to produce successively more abstract feature maps, corresponding to higher level properties of input data.

We observed that an attention layer was introduced into Seq2Seq systems to allow an LSTM encoder to track and adjust the associations among the elements of its context vectors and the components of the output vector of the decoder LSTM. Transformers dispense with recurrent processing entirely, and rely on multi-head attention units to drive their feed forward mechanisms. This permits them to achieve greater accuracy in learning, but it requires much larger pre-training on corpora for lexical embeddings. Transformers are converging on task general architectures that can be applied across a variety of applications, through either fast learning methods (GPT-3) or fine-tuning (BERT), for specific tasks.

We then considered Bizzoni and Lappin (2017)'s composite DL architecture for paraphrase assessment. This system combines LSTMs and CNNs in one encoder, and two such encoders in a classifier. It illustrates the use of different DNN elements within a single processing system in order to perform a challenging semantic task.

DNNs have become increasingly powerful through the use of multi-head attention, and large-scale pre-trained embeddings. This has facilitated transfer learning, where a DNN trained for one task can be easily adapted to others with the addition of fine-tuning layers. Through attention-driven architecture and pre-trained embeddings DNNs have come closer to domain general learning procedures that achieve a high level of performance across several domains, with limited task-specific training.

Over the past 15 years DNNs have moved from a niche technique of machine learning to the leading framework for work in AI. DL has achieved rapid progress across a wide range of AI tasks, approaching, and in some cases, surpassing human performance on these tasks. It has generated significant advances in several areas of NLP in which more traditional, symbolic methods have not yielded robust wide coverage systems after many years of work. These results are of cognitive interest to the extent that they show how it is, in principle, possible to effectively acquire certain types of linguistic knowledge through largely domain general learning devices.

CHAPTER 2

Learning Syntactic Structure with Deep Neural Networks

2.1 SUBJECT-VERB AGREEMENT

Both symbolic and statistical machine learning methods have been applied to syntactic learning tasks for many years, with varying levels of success. These include, *inter alia*, part of speech tagging, phrasal chunking, and sentential parsing.[1] DL has made significant progress across these, and other syntactic applications. Identifying subject-verb agreement is a particularly interesting application, because it involves long distance relations, and hierarchical structure.

Linzen, Dupoux, and Goldberg (2016) train an LSTM on a subset of a Wikipedia corpus to predict the number of a verb. As they observe, the task increases in difficulty in relation to the length of the sequence of NPs with the wrong number feature that occur between a subject and the verb that it controls. They refer to such intervening NPs as *attractors.* In 2.1 the subject-verb pairs are in italics, and the attractors are indicated in boldface.

2.1(a) *The students submit* a final project to complete the course.

(b) *The students* enrolled in **the program** *submit* a final project to complete the course.

[1] See A. Clark, Fox, and Lappin (2010) for discussions of both symbolic and statistical approaches to syntactic NLP tasks.

(c) *The students* enrolled in **the program** in **the Department** *submit* a final project to complete the course.

(d) *The students* enrolled in **the program** in **the Department** where **my colleague** teaches *submit* a final project to complete the course.

Linzen et al. (2016) use a dependency parser to identify the controlling subject of each verb in their corpus of examples. This identification is necessary to compute the number of attractors, but it is not used in the training for the number prediction task. The number of the verb is morphologically manifest in the raw data. They train their LSTM on ~121,500 examples (9% of the total corpus) by showing it the correct number feature of the verb. They test the LSTM's number predictions on ~1.21 million examples (90% of their corpus). They encode input words as vectors in 50 dimensions, and their LSTM has 50 hidden units. They report that their system achieves 99% accuracy in the number prediction task for cases with zero attractors between the subject and its verb. It declines to 83% accuracy for examples with four attractors. They do not report scores for examples with more than four attractors.

Linzen et al. (2016) also train a generative language model (giving a probability distribution to candidates for the next word in a sequence) to predict the number of the verb in an unsupervised manner. In contrast to their supervised LSTM, their language model scores below chance in its predictions for four attractor cases. The much larger Google LM (Józefowicz, Vinyals, Schuster, Shazeer, & Wu, 2016) does better, at a ~45% error rate for four attractors, but it is still well below their supervised RNN. They conclude that a DNN can learn a considerable amount of syntactic structure, if it is properly supervised.

2.2 ARCHITECTURE AND EXPERIMENTS

Bernardy and Lappin (2017) experiment with several DNN architectures, and alternative values for a variety of parameters, for the subject-verb agreement task. The architectures include an LSTM, a CNN, and a Gated Recurrent Unit (GRU) (Cho et al., 2014).

The parameters are

i. Ratio of training to testing as a partition of the corpus;

ii. Number of hidden units (memory size);

iii. Vocabulary size;

iv. Number of layers;

v. Dropout rate; and

vi. Lexical embedding dimension size.

A GRU network is an RNN whose processing architecture differs from an LSTM in several of its filtering and update procedures. The design of a GRU is shown in Fig 2.1.

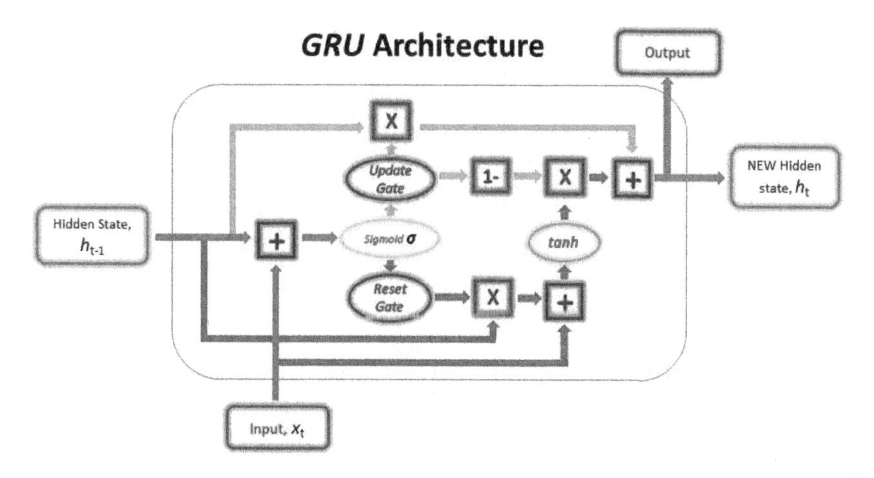

Figure 2.1 GRU.
From Gabriel Loye, "Gated Recurrent Unit (GRU) With PyTorch", *FloydHub*, July 22, 2019.

Bernardy and Lappin (2017)'s CNN has 6 levels, with filtering successively compressing vector dimensions from 50, through 20, 15, 10, to 5. Convolution at these levels yields 7, 5, 5, and 3 features, respectively. Every convolution layer uses a ReLU activation function.[2] The last layer is dense, with sigmoid activation. Fig 2.2 displays its structure.

[2]A Rectified Linear Activation (ReLU) function returns its input argument if it is a positive value, and 0 otherwise. It is specified by the equation

$$f(x) = max(0, x).$$

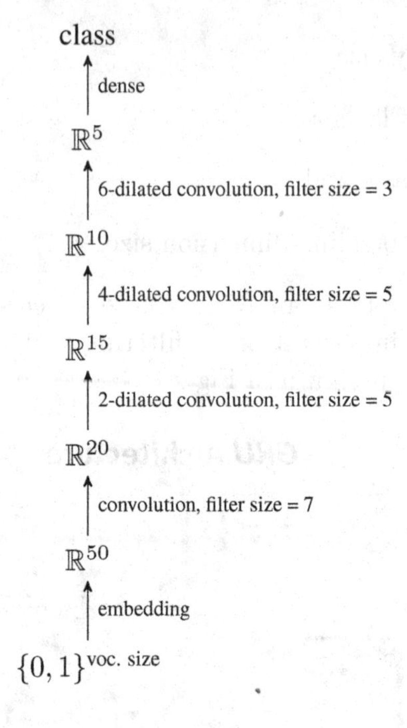

Figure 2.2 Bernardy and Lappin (2017)'s CNN.

Bernardy and Lappin (2017) use the WaCkypedia English corpus (Baroni, Bernardini, Ferraresi, & Zanchetta, 2009), which contains ~24 million example cases of present tense subject-verb agreement. The corpus is annotated with POS tags by TreeTagger (Schmid, 1995), and with dependency relations by the MaltParser (Nivre et al., 2007). Linzen et al. (2016) restrict training and testing to one agreement case per sentence in their corpus. Bernardy and Lappin (2017) use the full set of number agreement relations in the sentences of their corpus.

Linzen et al. (2016) limit their test, but not their training examples to cases in which all NPs intervening between the subject and the verb that it controls are attractors. Bernardy and Lappin (2017) include the cases in which agreeing, as well as non-agreeing NPs intervene between the subject and its verb. Their motivation for departing from Linzen et al. (2016)'s experimental design is a concern to measure the accuracy with which a DNN predicts verb number in complex, and possibly confusing, syntactic sequences.

Bernardy and Lappin (2017) first identify a benchmark of reasonable performance for the supervised DNN configuration and training. The benchmark is an LSTM with one layer of 150 units and no dropout, a dataset constructed with 10,000 words, lexical embeddings of dimension 50, and a training regimen of 90% of the corpus. They then run experiments varying each of these parameters independently, holding the others constant. They experiment with the following parameter values for their DNNs:

- Training on 10%, 50%, and 90% of the corpus, testing on the remainder for each split;
- 50, 150, 450, and 1350 units for the LSTM layers;
- Embedding vocabulary sizes of 100, 10k, and 100k words, substituting corresponding POS tags for the rest;
- 1, 2, and 4 layers for the LSTM;
- Dropout rates of 0, 0.1, 0.2, and 0.5, applied to the weights within the LSTM layers, but not at the final dense layer; and
- Lexical embedding dimension sizes of 17, 50, 150, and 450.

They hypothesise that a DNN will learn the target syntactic dependency pattern more efficiently if it is exposed to input consisting largely of POS sequences in which number features are marked on noun and verb tags. They conjecture that such input would highlight the dependency relations more clearly by abstracting away from possibly confounding distributional lexical information contained in richer embeddings. On this view impoverished lexical sequences would facilitate DNN learning of agreement patterns through highlighting the relevant number feature.

Bernardy and Lappin (2017) also train an LSTM as a generative language model. The LSTM has two layers of 1200 units per layer, and a dropout rate of 0.5. It is trained on the WaCky corpus of sentences, with the 100 most common words, and corresponding POS tags substituted for the others in the sentences. This design is intended to test the reduced vocabulary hypothesis. They use their neural LM for unsupervised prediction of agreement in two ways. Let $p(w_i|w_{i-1}, ..., w_{i-k})$ be the predicted probability of a word w in a string, given the prefix of $w_{i-1}, ..., w_{i-k}$ of preceding words in that string. On the first approach, they determine, for each sentence in the test set, whether the following condition holds:

$$\sum_{V^n} p(V_i^n | w_{i-1}, ..., w_{i-k}) > \sum_{V^{\neg n}} p(V_i^{\neg n} | w_{i-1}, ..., w_{i-k})$$

where V^n ranges over verbs with the correct number feature (n) in a particular string, and $V^{\neg n}$ ranges over verbs with the wrong number feature in that string. On the second method they test to see if the following condition holds:

$$p(Verb_i^n | w_{i-1}, ..., w_{i-k}) > p(Verb_i^{\neg n} | w_{i-1}, ..., w_{i-k})$$

The first method measures the summed conditional probabilities of all correctly number inflected verbs after the prefix in a string against the summed predicted probabilities of all incorrectly number inflected verbs appearing in this position (*the summing method*). The second procedure compares the predicted probability of the correctly number-marked form of the actual verb in this position with that of its incorrectly marked form (*the verb targeted method*). While in the summing method the LM is not given any semantic cue, in the verb targeted method it is primed to expect a specific verb. This bias contributes minimal semantic information, given that the Bernardy and Lappin (2017) LM's 100-word vocabulary contains only the inflectional verb forms *to be*, *to have*, and *to state*, with other verbs represented by the **VV** part of speech code.

Bernardy and Lappin (2017) present the results of their experiments in a series of graphs, which I reproduce here. In Figs 2.3–2.11 the y-axis gives the accuracy rate, and the x-axis the number of NP attractors (from 0 to 7). 73% of the verbs in the test sets are singular. This provides a majority choice baseline, which is indicated by a straight horizontal line in the graphs.

Figs 2.3 and 2.4 show that a reduced vocabulary of the 100 most common words, with POS tags for the remainder, degrades accuracy across DNN architectures, for supervised learning. Increasing the vocabulary to 100k words yields a significant improvement for the LSTM, but gives mixed results for the CNN.

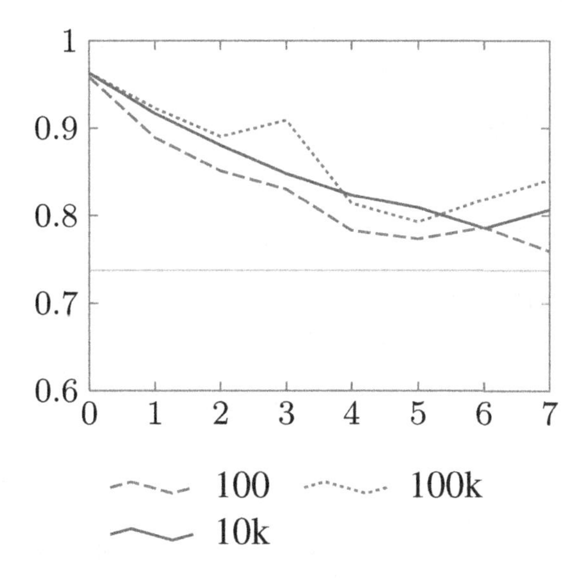

Figure 2.3 Accuracy of the supervised LSTM relative to vocabulary size.

Fig 2.5 indicates that increasing the ratio of training to testing examples from 10% to 50% significantly improves the performance of the LSTM (with 150 units and a vocabulary of 10,000 word embeddings). Further increasing it to 90% does not make much of a difference, even degrading accuracy slightly at six attractors.

We see from Fig 2.6 that the LSTM and GRU perform at a comparable level, and both achieve significantly better accuracy than the CNN.

Fig 2.7 shows that increasing the number of units in an LSTM improves accuracy relative to the number of attractors. Each three-fold increase in units achieves a similar improvement in percentage points for a higher number of attractors, up to 450 units.

Increasing the number of layers for an LSTM from 1 150-unit layer to 2 such layers marginally improves its performance (Fig 2.8). A further increase to 4 150-unit layers makes no clear difference.

A dropout rate of 0.1 for the LSTM RNN (1 layer with 150 units) improves LSTM performance slightly (Fig 2.9). An increase to 0.2 provides no clear benefit, while increasing the rate to 0.5 degrades performance.

The DNN configured with the best observed parameter values is an LSTM with 2 layers, 1350 units, a dropout rate of 0.1, a vocabulary size

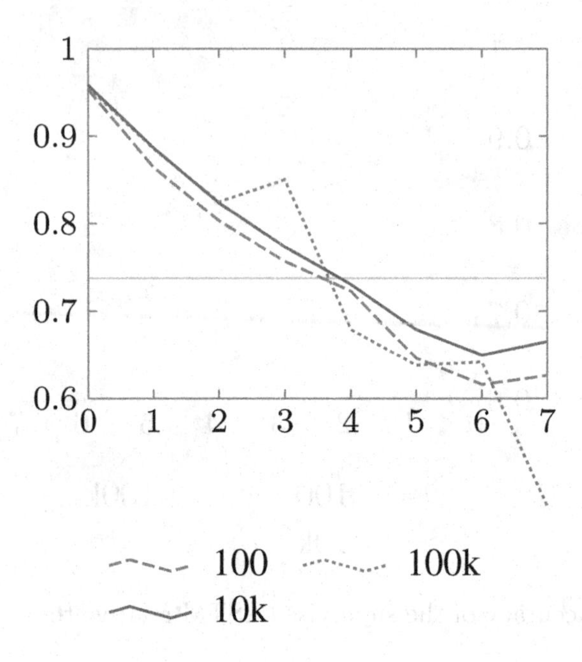

Figure 2.4 Accuracy of the CNN relative to vocabulary size.

of 100k, trained on 90% of the corpus, and lexical embedding size of 150 dimensions (Fig 2.10).

For the unsupervised LSTM LM the verb targeted method achieves a far higher level of accuracy than Linzen et al. (2016)'s LM, and the much larger Google LM. However, it is still below the best LSTM supervised results (Fig 2.11).

There is an inverse relation between the number of examples in the corpus and the number of attractor NPs in these sentences (2.12).

Bernardy and Lappin (2017)'s results support Linzen et al. (2016)'s finding that RNNs (both LSTMs and GRUs) learn long distance syntactic dependencies within extended, complex sequences. Their success in learning subject-verb agreement scales with the size of the dataset on which they are trained. Training DNNs on data that is lexically impoverished, but highlights the syntactic elements of a relation, does not (for this task) facilitate learning, but degrades it, contrary to their initial hypothesis. This suggests that DNNs extract syntactic patterns incrementally from lexical embeddings, through recognition of their distributional

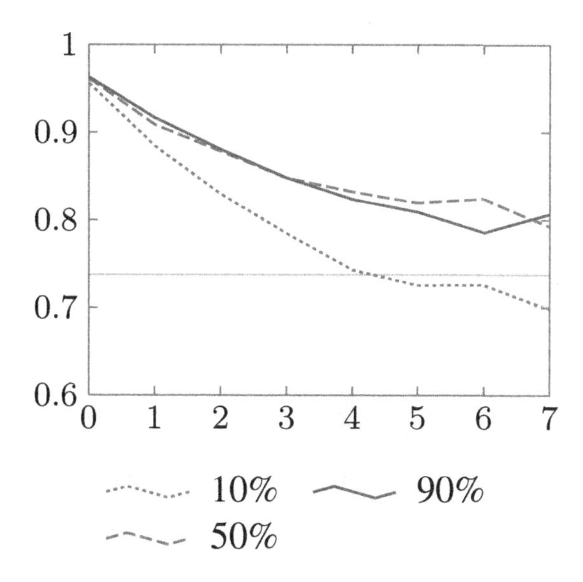

Figure 2.5 Influence of the size of the training corpus on LSTM accuracy.

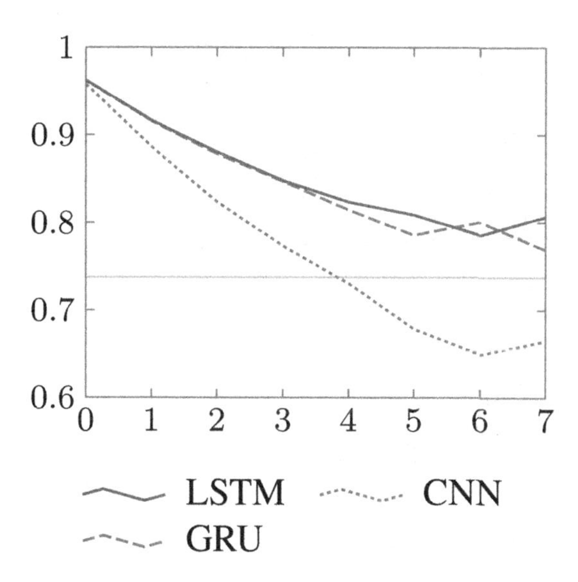

Figure 2.6 Relative accuracy of the LSTM, GRU, and CNN.

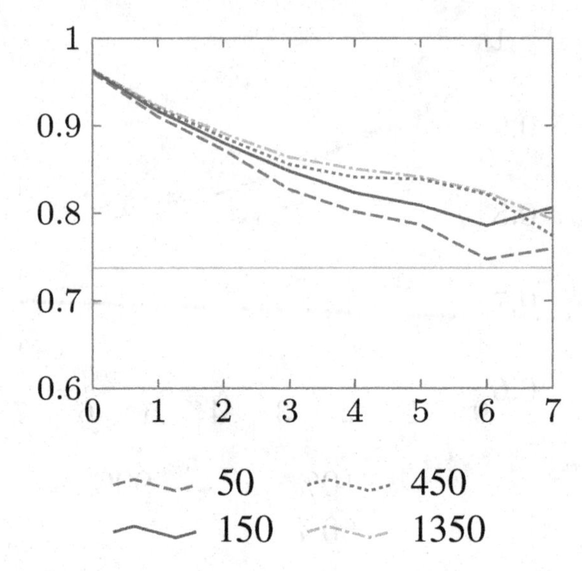

Figure 2.7 LSTM accuracy relative to memory size.

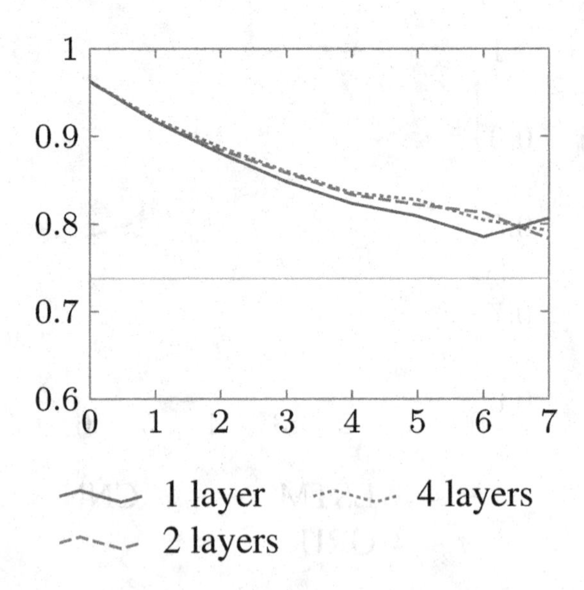

Figure 2.8 LSTM accuracy relative to layers.

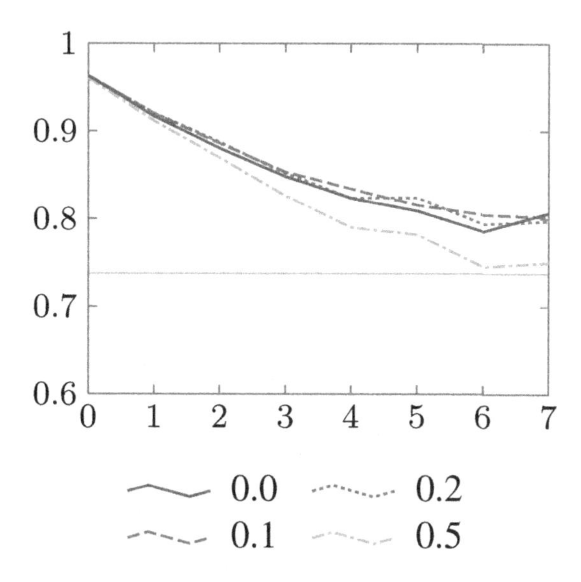

Figure 2.9 LSTM accuracy relative to dropout rate.

regularities. Their results also show that an unsupervised language model can achieve reasonable results on the agreement prediction task. It is important to note that their LM is lexically impoverished, and it is likely that this undermines its performance. As we will see later in this chapter, more recent work shows that LSTM LMs with larger vocabularies can perform at or above the level of accuracy that the supervised LSTM achieves.

It is an open question as to how DNN learning resembles and diverges from human learning. Bernardy and Lappin (2017) make no cognitive claims concerning the relevance of their experiments to human syntactic representation. It is interesting to note that some recent work in neurolinguistics indicates that syntactic knowledge is distributed through different language centres in the brain, and closely integrated with lexical-semantic representations (Blank, Balewski, Mahowald, & Fedorenko, 2016). This lexically encoded and distributed way of representing syntactic information is consistent with the role of rich lexical embeddings in DNN syntactic learning. The results showing a clear correlation between vocabulary size and the accuracy of the supervised LSTM subject-verb identification supports this analogy.

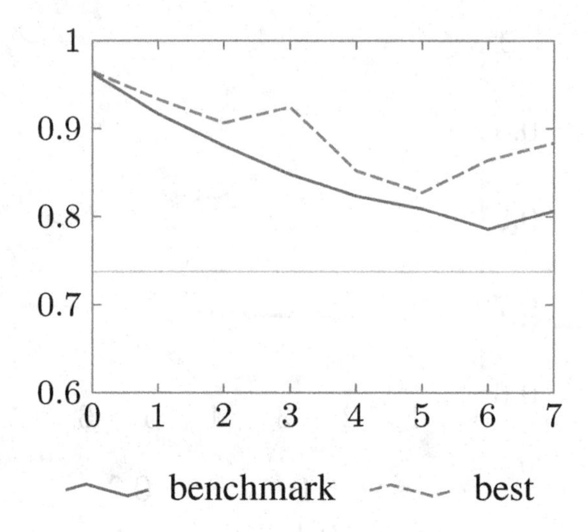

Figure 2.10 Optimised LSTM performance relative to the benchmark LSTM.

In more recent work, Schrimpf et al. (2020) show that transformer models, specifically GPT-2 and BERT, accurately predict human brain responses to linguistic input, over a variety of language processing tasks. Much additional work in both neuroscience and DNN modelling of linguistic knowledge is required before we can draw substantive conclusions on this issue. But apparent parallels between human and DNN syntactic representation are intriguing, and worth exploring further.

2.3 HIERARCHICAL STRUCTURE

Gulordava, Bojanowski, Grave, Linzen, and Baroni (2018) use an LSTM LM for unsupervised learning of subject-verb agreement, with tests sets in Italian, English, Hebrew, and Russian. They extract test sentences from a dependency tree bank, and convert them into nonsense (nonce) sentences by substituting arbitrary lexical items with corresponding POS, from the LM's vocabulary, for the nouns and verbs in the originals. The test set is comprised of both the original and the nonsense sentences. They train the LSTM LM on Wikipedia text from the four languages, with 50m tokens in each training corpus.

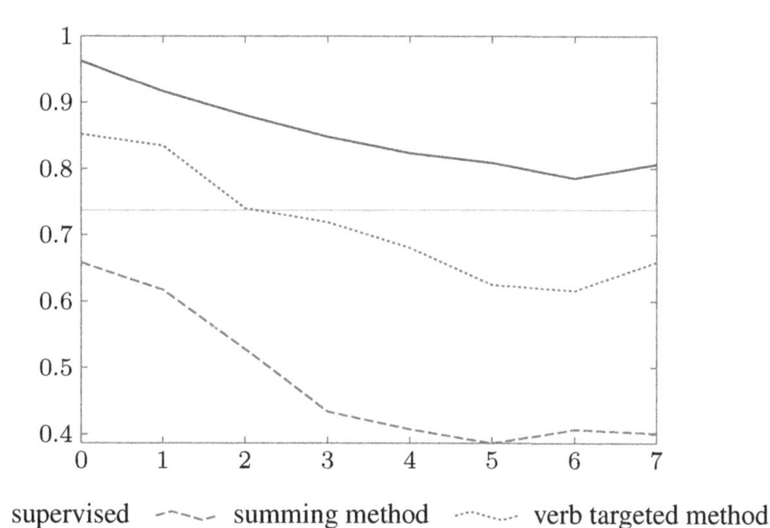

Figure 2.11 Unsupervised LSTM LM model relative to the supervised LSTM.

Gulordava et al. (2018) test their LSTM LM for accuracy and model perplexity against three baseline systems.[3] These include a unigram majority baseline model, a 5-ngram model with Kneser-Ney smoothing, and an LSTM limited to a window of five preceding words in a sequence.[4] Their results, for both original and nonce test sentences, are given in Table 2.1.

---

[3]The perplexity of a language model measures the degree of uncertainty in the likelihood of the probability values that it assigns to words and sequences of words. Lower perplexity correlates with higher certainty in predicting sentence probability. Perplexity is the logarithm of entropy. Therefore the following equation holds:

$$H(P, Q) = log\ PP(P, Q)$$

where $H(P, Q)$ is the entropy of the probability distribution Q relative to the distribution P (cross entropy, or loss, of Q relative to P), and $PP(P, Q)$ is the perplexity of Q relative to P.

[4]Kneser-Ney smoothing is a procedure for estimating the probability distribution of an ngram relative to its history in a document. See Goodman (2001) for interpolated Kneser-Ney smoothing in ngram language modelling.

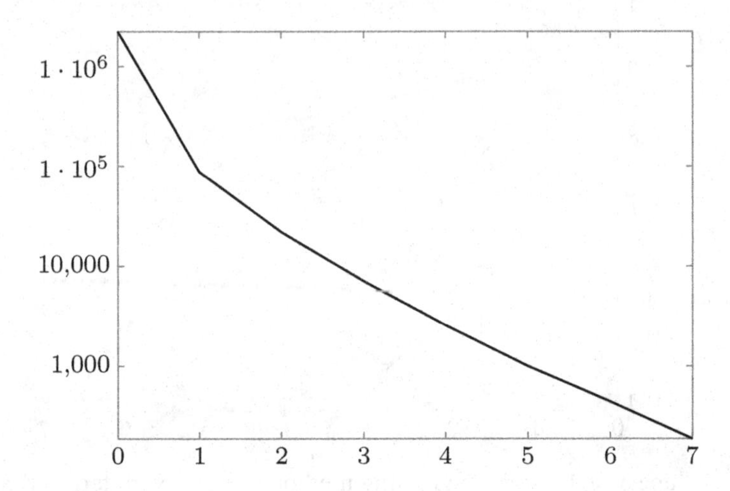

Figure 2.12 Ratio of examples in the corpus to attractors in a sentence.

Gulordava et al. (2018)'s LM significantly outperforms the three baseline systems. The fact that it yields reasonable accuracy for nonce sentences indicates that it learns hierarchical syntactic structure independently of semantic cues. But its predictive accuracy is still significantly higher for the original sentences, showing that these cues facilitate syntactic learning, as Bernardy and Lappin (2017)'s work suggests. The model performs better with morphologically richer languages, like Hebrew, and Russian, than with English. There is a correlation between the quality of the LM (perplexity value) and its accuracy on the agreement task.

They also compare the performance of their LSTM LM to Amazon Mechanical Turk (AMT) crowd-sourced human predictions for their Italian dataset. They report that the LSTM model approaches the human level of performance. For the original sentences, average human accuracy is 94.5, and the LSTM LM is 92.1. For the nonce sentences, average human accuracy is 88.4, and the LSTM LM is 85.5. These results provide additional support for the conjecture that there are parallels in the ways that humans and DNNs learn and represent certain types of syntactic information.

Marvin and Linzen (2018) test an LSTM LM on subject-verb agreement in ten different syntactic structures. They present the model with

TABLE 2.1 Accuracy and Perplexity of Guldorova et al. (2018)'s LSTM LM

	Italian	English	Hebrew	Russian
Unigram Majority Baseline				
Original	54.6	65.9	67.8	60.2
Nonce	54.1	42.5	63.1	54.0
5-ngram Kneser-Ney smoothing				
Original	63.9	63.4	72.1	73.5
Nonce	52.8	43.4	61.7	56.8
Perplexity	147.8	168.9	122.0	166.6
5-gram LSTM				
Original	81.8	70.2	90.9	91.5
Nonce	78.0	58.2	77.5	85.7
Perplexity	62.6	71.6	59.9	61.1
LSTM				
Original	92.1	81.0	94.7	96.1
Nonce	85.5	74.1	80.8	88.8
Perplexity	45.2	52.1	42.5	48.9

minimal pairs of correct and ill-formed instances of each structure, rating the LM's prediction as accurate if it assigns higher probability to the well-formed sentence. They also use AMT crowdsourcing to obtain binary human judgements on these pairs. In 5 of the 10 constructions the LSTM LM approaches or surpasses human performance, but in the other 5 it scores 21–35 points below the human judgement standard. This result sheds some doubt on our conjecture.

However, Goldberg (2019) provides additional experimental work on the Marvin and Linzen (2018) test set. He uses BERT (Base and Large, both untuned) as a LM on this set. He masks the contrasting verbs in each minimal pair, and uses BERT's pre-softmax logit scores as quasi-probability values to test its prediction for each of these verbs.[5]

By virtue of its bidirectional processing of input, BERT conditions its predicted scores on both the left and right contexts of the masked verb. In 9 of the 10 syntactic constructions, BERT approaches or surpasses Marvin and Linzen's reported human judgement accuracy, in some cases by a significant margin. In the one exception, sentential complements, BERT Base scores 0.83, and BERT Large 0.86, while human accuracy is 0.93.

Kuncoro et al. (2018) introduce an architectural syntactic bias into the design of an LSTM LM, for the subject verb recognition task. They apply Recurrent Neural Network Grammars (RNNGs) (Dyer, Kuncoro, Ballesteros, & Smith, 2016) to Linzen et al. (2016)'s test set. Their RNNG LM contains a stack and a composition operator for predicting the constituency structure of a string. It assigns joint probability to a string and a phrase structure. This RNNG and their best sequential LSTM LM achieve approximately the same level of accuracy in predicting agreement for sentences with 0-2 attractors (98%, 96.5%, and 95%, respectively). However, the RNNG outperforms the LSTM LM for cases with higher numbers of attractors, with 93% vs 87% at 4 attractors, and 88% vs 82% at 5 attractors).

An RNNG is designed to induce hierarchical structure on a string. The fact that it outperforms an LSTM lacking this architectural feature, on the agreement task, for more complex cases, suggests that it is better

[5]Because BERT predicts words on the basis of their left and right contexts, it is not possible to compute the conditional probability that Bert assigns to a word in a sequence solely on the basis of the probabilities of the words that precede it. Therefore, we cannot use the formula that unidirectional LMs apply to obtain the probability of a sentence,

$$P(s) = \prod_{i=0}^{|s|} P(w_i | w_{<i}).$$

Instead, we require the bidirectional probability formula

$$P(s) = \prod_{i=0}^{|s|} P(w_i | w_{<i}, w_{>i}).$$

This equation does not yield true probabilities, as its values do not to sum to 1. Normalising these values to genuine probabilities through dividing by the set of all possible sentences is intractable. Therefore, it is necessary to use quasi-probabilities as confidence scores, in the way that Goldberg does. See also Kuncoro et al. (2020) on this issue. I return to it in Chapter 4.

able to learn long distance syntactic dependencies. However, Kuncoro et al. only test their RNNG LM on a limited set of English data. Gulordava et al. (2018)'s experiments indicate that simple LSTMs achieve very high accuracy on the agreement task, for morphologically richer languages, and they replicate human performance for Italian.

Kuncoro, Dyer, Rimell, Clark, and Blunsom (2019) observe that RNNGs do not scale up to large training sets, because of the high computational cost of training them to predict both strings and parse structures for large corpora. As an alternative they use an RNNG as a teacher model to supervise the training of a sequential LSTM over large dataset. They describe this process as knowledge distillation. It appears to induce the RNNG's hierarchical structural bias in the sequential LSTM LM's predictions. Specifically, through RNNG mentoring in training, the sequential LSTM LM limits its probability distributions to those that the RNNG would produce for sequences.

Kuncoro et al. (2019) report that their Distilled Syntax-Aware LSTM (DSA-LSTM) scores, on average, 90% on Gulordava et al.'s English subject-verb agreement test set. Their non-DSA-LSTM obtains 87%, and BERT 89%. These scores are comparable, and they are all slightly above average human performance for this test set, which they give as 86%. This suggests that, with sufficient training data, both sequential LSTMs and bidirectional transformers can achieve high accuracy on the agreement task, even without prior hierarchical structure bias.

2.4 TREE DNNS

TreeRNNs (Bowman et al., 2016; Socher, Pennington, Huang, Ng, & Manning, 2011) are trained to assign syntactic trees to input sentences by supervised learning on parse structure annotations. These models have achieved improved performance on tasks like natural language inference and sentiment analysis.

In NLI, a DNN is trained on a labelled corpus of inferences from premises to conclusions, to classify unlabelled inferences in a test set. Each such inference is annotated as an *entailement*, *neutral* (contingent), or a *contradiction*. The inferences in 2.2 are examples from the development part of Bowman, Angeli, Potts, and Manning (2015)'s Stanford NLI corpus, where the gold labels are given in bold.

2.2(a) A man inspects the uniform of a figure in some East Asian country. →
The man is sleeping. **contradiction**

(b) An older and younger man smiling. →
Two men are smiling and laughing at the cats playing on the floor. **neutral**

(c) A black race car starts up in front of a crowd of people. →
A man is driving down a lonely road. **contradiction**

(d) A soccer game with multiple males playing. →
Some men are playing a sport. **entailment**

(e) A smiling costumed woman is holding an umbrella. → A happy woman in a fairy costume holds an umbrella. **neutral**

For sentiment analysis one trains a DNN on a corpus in which phrases, sentences, or articles are annotated with sentiment labels. The DNN is tested on an unlabelled set for accuracy of sentiment classification. So, for example, Socher et al. (2011) annotate a corpus of stories with the following five category labels (in bold), where their interpretation of each category is indicated in italics.

- 1. **Sorry, Hugs**: *User offers condolences to author*;
- 2. **You Rock**: *Indicating approval, congratulations*;
- 3. **Teehee**: *User found the anecdote amusing*;
- 4. **I Understand**: *Show of empathy*;
- 5. **Wow, Just Wow**: *Expression of surprise/shock*.

Latent tree RNNs (Choi, Yoo, & Lee, 2018; Maillard, Clark, & Yogatama, 2019; Yogatama, Blunsom, Dyer, Grefenstette, & Ling, 2017) learn to induce tree representations without supervised learning on parse annotations. Williams, Drozdov, and Bowman (2018) test several TreeRNNs, supervised and latent, as well as a baseline non-tree LSTM, on two NLI test sets: the Stanford NLI set, and the MultiNLI corpora (Williams, Nangia, & Bowman, 2018).

Choi et al. (2018)'s latent tree RNN (ST-Gumbel) outperformed all other systems, scoring accuracy rates of 83.7 on SNLI, and 69.5 on

MultiNLI. The other systems scored between 81.3 and 82.6 on SNLI, and between 66.2 and 69.1 on MNLI. The non-tree LSTM achieved the second highest rate of accuracy, with 82.6 on SNLI, and 69.1 on MNLI. Williams, et al. (2018) observe that Choi et al.'s ST-Gumbel model, and the other latent tree RNNs that they test, yield shallow parse structures, which are not consistent across test sentences.

Williams, et al. (2018) also point out that the elements of these parses do not correspond to the constituents of either the Penn Tree Bank (PTB) (M. P. Marcus, Santorini, & Marcinkiewicz, 1993), or those of formal syntactic theories. Given these results, and the fact that the non-tree LSTM baseline achieves results comparable to Choi et al. (2018)'s, it is unclear to what extent tree representations, latent or supervised, contribute to the NLI task. It is necessary to test other DNN architectures, both with and without parse representations, across a variety of NLP tasks, to determine whether tree structures enhance performance on these tasks.

Hewitt and Manning (2019) use a supervised probe to test for the presence of implicit syntactic tree structure in the word vectors of DNNs. The probe consists of a squared L2 distance measure, which determines the distance (number of nodes) between word pairs in a tree, and a squared L2 norm, which identifies the relative depth (chain of nodes from the root) of each word in a tree.[6] This metric defines a mapping from the word vectors in the sentence representation of a DNN into a parse tree structure. Hewitt and Manning (2019) test their probe on the word representations of the unlabelled Stanford Dependency parses (de Marneffe, MacCartney, & Manning, 2006) for the parsing train/dev/test splits of the PTB.

Dependency parsers represent the syntactic structure of a phrase as a hierarchical graph of heads and dependents, with the main verb constituting the root of the phrase. Fig 2.13 shows the labelled dependency parse tree for the sentence *This time around, they're moving even faster.*

Hewitt and Manning (2019) find that ELMO and BERT significantly outperform baseline systems in predicting unlabelled dependency parse trees for the PTB training set. ELM01 scores an average Spearman correlation of 0.87 on word distances, and 0.87 on word depth.[7] $\text{BERT}_{LARGE15}$ achieves an average Spearman correlation of 0.87 on distance, and 0.89

[6]For a vector v, the squared L2 norm of v is the square root of the sum of the squared values of the elements of v.

[7]See Chapter 1, fn 7 for brief explanations of the Spearman and Pearson correlation metrics.

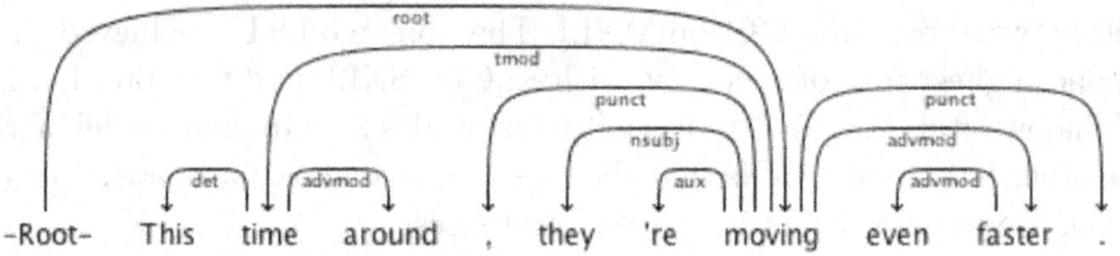

Figure 2.13 Stanford Labelled Dependency Parse Tree.
From https://nlp.stanford.edu/software/nndep.html

on depth. These scores express the degree of correspondence between the unlabelled dependency parses that the probe identifies in the sentential output vectors which ELMO and BERT produce on one hand, and the gold standard dependency parses (without their labels) for these sentences in the PTB on the other. The results suggest that the lexical embeddings of large pre-trained transformers incorporate consistent hierarchical syntactic information corresponding to dependency parse trees.

2.5 SUMMARY AND CONCLUSIONS

In this chapter, we have considered recent experimental work on the application of DNNs to several NLP tasks that involve learning to identify hierarchical syntactic structure. We started with subject-verb agreement, looking at the performance of both supervised LSTMs and unsupervised LSTM language models. We then briefly reviewed work on more refined test sets for agreement across a variety of syntactic constructions. BERT shows a significant improvement in accuracy over LSTM LMs on this test set, approaching or surpassing human performance on almost all syntactic categories.

We next took up LSTM models that incorporate hierarchical syntactic bias in their design, or in their training regimen. In general, non-biased sequential LSTM LMs do comparably well, if provided with sufficient training data. We concluded with treeRNNs, and a tree probe applied to transformers. The former did not seem to offer a significant advantage over non-tree LSTMs for the tasks to which they were applied.

The tree probe identifies reliable dependency tree structures in the lexical embeddings of ELMO and BERT, but not in non-transformer DNNs.

The experimental work considered here strongly indicates that DNNs learn a considerable amount of hierarchical syntactic structure. This information is implicit in the distributed lexical embeddings of a DNN, and it is encoded in the vector representations of the words, phrases, and sentences that the DNN produces. Therefore, training (or pre-training) a DNN on a large vocabulary seems to be a necessary condition for syntactic learning.

The work that we explored in this chapter raises two important questions. First, will incorporating structural bias into a DNN improve its performance on linguistically interesting applications, and possibly allow for a substantial reduction in required training? Second, to what extent does the distributed, lexically driven nature of deep syntactic learning resemble the way in which humans acquire and represent knowledge of the syntactic properties of their language? The experimental evidence that we reviewed in this Chapter indicates that prior syntactic bias does not significantly improve the performance of a DNN, at least for the tasks that we considered here. However, this is a tentative and preliminary conclusion. We need to look at work on other tasks, with different sorts of DNN architectures. We will do this in the following chapters. Some initial neurolinguistic results that we cited suggests a positive answer to the second question. However, this remains a conjecture. Again, a great deal of additional research both on the neurological foundations of human language acquisition and linguistic representations, and on the nature of deep learning, is needed in order to have confidence in this conjecture. Both questions remain tantalisingly open.

In considering DNN performance on an NLP task one of our main concerns should be to determine to what extent the model can be applied to other, entirely different tasks, with minimal (optimally, no) modification of design, and limited training. If we are to sustain the view that DNNs provide largely domain general inductive learning devices, then it should not be necessary to reconfigure them for each task. In fact one of the main advantages of large pre-trained transformers like BERT is that they easily accommodate general purpose architecture, with transfer learning among tasks. They can be adapted to a new application through fine-tuning, which involves small-scale supplementary training. However, as we saw in our brief discussion of GPT-3 in Chapter 1, robust transfer learning and minimal domain–specific instruction may come at the cost of massive pre-training. These are core issues in deep learning, particularly in its application to NLP. While they are obviously major engineering considerations in the development of efficient processing systems, they also have cognitive significance. As we noted in Chapter 1, the extent to which a domain general learning system can achieve human level performance in a linguistically interesting task indicates how the knowledge required for this task could, in principle, be acquired without strong domain specific learning biases.

CHAPTER 3

Machine Learning and the Sentence Acceptability Task

3.1 GRADIENCE IN SENTENCE ACCEPTABILITY

The use of neural language models in NLP tasks that require syntactic knowledge raises the question of the relationship between grammaticality and probability. A. Clark and Lappin (2011) and Lau, Clark, and Lappin (2017) show that grammaticality cannot be directly reduced to probability. Such a reduction requires specifying a threshold probability value k such that only sentences with a probability $\geq k$ are grammatical. Such a threshold entails that the cardinality of the set of grammatical sentences is finite, which is not the case for any natural language. Assume, for example, that given a language model M for a set of sentences S (finite or infinite), only sentences in S with probability 0.1 or higher in M are grammatical. The probability distribution that M generates allows only for a finite $S' \subseteq S$ of sentences in which, for any $s' \in S', P_M(s') \geq 0.1$ (in this case $|S'| \leq 10$). Clearly this result holds for any choice of M, S, and threshold probability value.

Grammaticality is a theoretical property, which is not directly accessible to observation. Speakers' acceptability judgements can be observed and measured. These judgements provide the primary data for most linguistic theories. An adequate theory of syntactic knowledge must be able to account for the observed data of acceptability judgements. The experimental work that I discuss in this chapter, and the following one, measures and predicts speakers' judgements on the acceptability of sentences. These are elicited through AMT crowd-source experiments in which annotators rate sentences in an AMT Human Intelligence Task (HIT) for naturalness. Sentence acceptability provides evidence for the grammatical status of a sentence.

The same argument that A. Clark and Lappin (2011) and Lau, Clark, and Lappin (2017) use to show that grammaticality cannot be reduced directly to a probability value threshold applies to acceptability. However, it is possible to construct models in which probability distributions provide the basis for predicting relative acceptability, and through it, degree of grammaticality. I will explore these models in the next two chapters.

Lau, Clark, and Lappin (2014, 2015, 2017) (LCL) present extensive experimental evidence for gradience in human sentence acceptability judgements. They show that crowd-sourced judgements on round-trip machine translated sentences from the British National Corpus (BNC) exhibit both aggregate (mean) and individual gradience. They use Google Translate to map the sentences in their test sets into four target languages, Norwegian, Spanish, Chinese, or Japanese, and then back into English. The purpose of round-trip MT is to introduce a wide variety of infelicities into some of the sentences, to ensure variation in acceptability judgements across the examples of the set. Each HIT contains one original, non-translated sentence, which is used to control for annotator fluency. LCL test three modes of presentation of their HITS: binary, four categories of naturalness, and a sliding scale with 100 underlying points. They find a high Pearson correlation ($\geq$0.92) between the three modes of presentation, and so they adopt the four category HIT format for subsequent experiments. Fig 3.1 displays a four category sentence acceptability rating HIT.

Figs 3.2 and 3.3 give the histograms of mean acceptability ratings in four category and slider modes of presentation, respectively. Figs 3.4 and 3.5 show the histograms for individual four category and slider ratings, respectively.

Lau et al. (2015) and Lau, Clark, and Lappin (2017) also demonstrate that crowd-sourced judgements on linguists' examples from Adger (2003), in which semantic/pragmatic anomaly has been filtered out, display the same sort of gradience for both mean and individual acceptability ratings. Figs 3.6 and 3.7 display the mean acceptability ratings (with four category presentation) for the good and the starred sentences

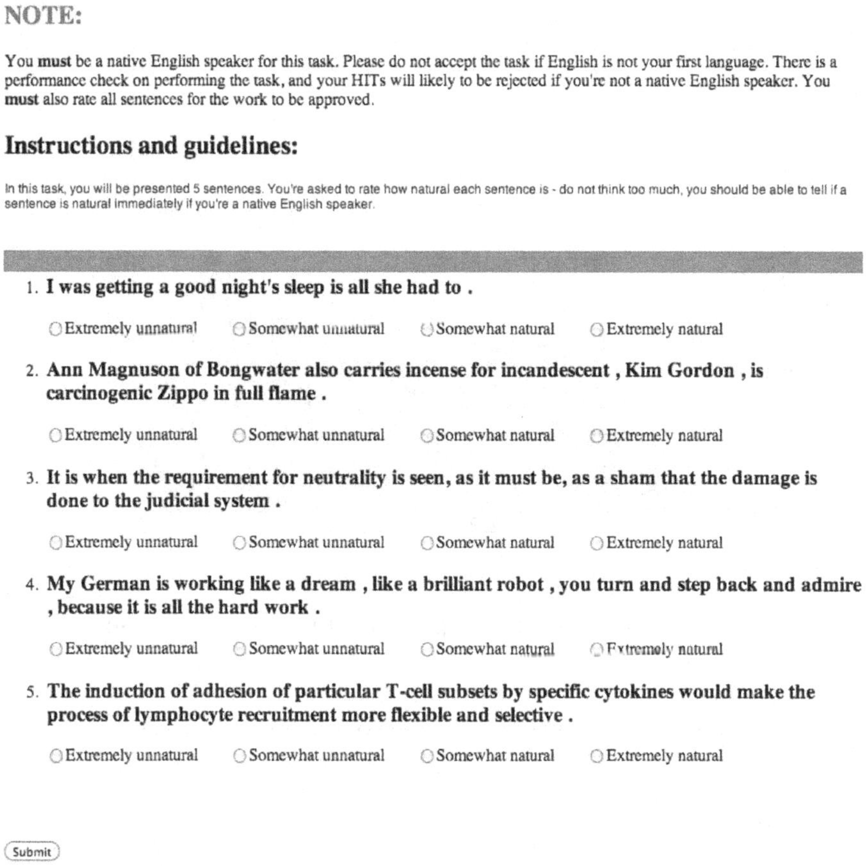

NOTE:

You **must** be a native English speaker for this task. Please do not accept the task if English is not your first language. There is a performance check on performing the task, and your HITs will likely to be rejected if you're not a native English speaker. You **must** also rate all sentences for the work to be approved.

Instructions and guidelines:

In this task, you will be presented 5 sentences. You're asked to rate how natural each sentence is - do not think too much, you should be able to tell if a sentence is natural immediately if you're a native English speaker.

1. **I was getting a good night's sleep is all she had to .**

   ○ Extremely unnatural ○ Somewhat unnatural ○ Somewhat natural ○ Extremely natural

2. **Ann Magnuson of Bongwater also carries incense for incandescent , Kim Gordon , is carcinogenic Zippo in full flame .**

   ○ Extremely unnatural ○ Somewhat unnatural ○ Somewhat natural ○ Extremely natural

3. **It is when the requirement for neutrality is seen, as it must be, as a sham that the damage is done to the judicial system .**

   ○ Extremely unnatural ○ Somewhat unnatural ○ Somewhat natural ○ Extremely natural

4. **My German is working like a dream , like a brilliant robot , you turn and step back and admire , because it is all the hard work .**

   ○ Extremely unnatural ○ Somewhat unnatural ○ Somewhat natural ○ Extremely natural

5. **The induction of adhesion of particular T-cell subsets by specific cytokines would make the process of lymphocyte recruitment more flexible and selective .**

   ○ Extremely unnatural ○ Somewhat unnatural ○ Somewhat natural ○ Extremely natural

Submit

Figure 3.1 Four category AMT sentence acceptability HIT.

in Adger (2003), while Figs 3.8 and 3.9 show the individual ratings for these examples.

One might think that humans have a tendency to treat all classifiers as gradient. In fact, this is not the case. Lau et al. (2015) and Lau, Clark, and Lappin (2017) experiment with non-linguistic classifiers. They show that while AMT annotators judge body weight as gradient in drawings of human figures (Fig 3.10), their judgements of even vs odd natural numbers are sharply binary (Fig 3.11).

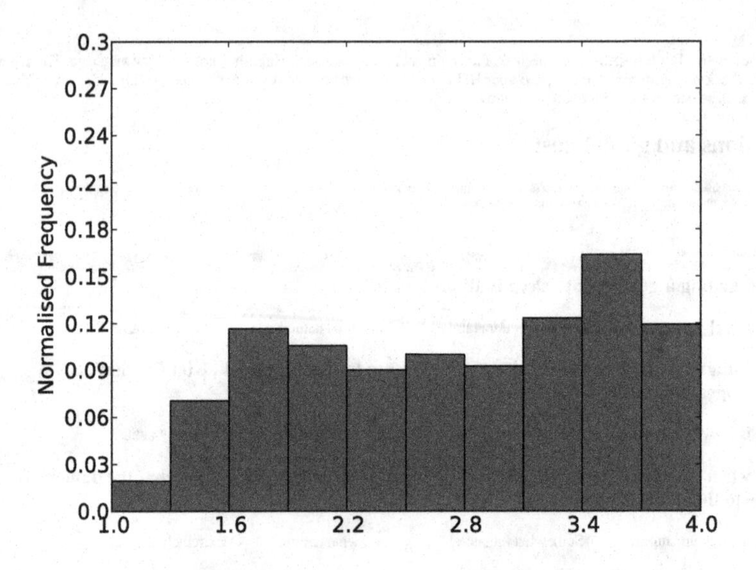

Figure 3.2 Four category mean acceptability ratings for BNC sentences.

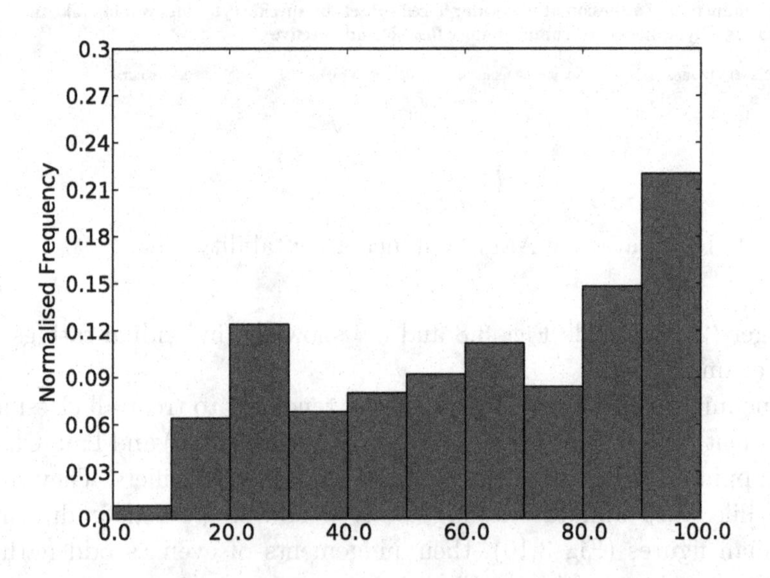

Figure 3.3 Slider mean acceptability ratings for BNC sentences.

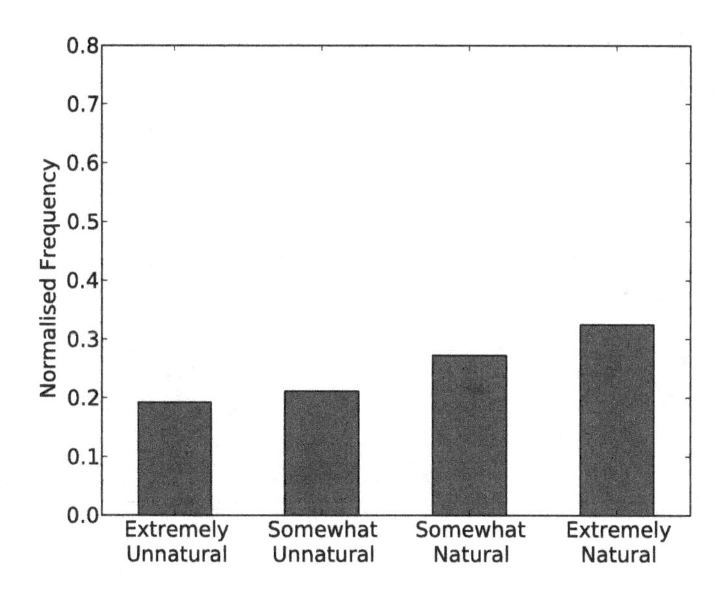

Figure 3.4 Four category individual acceptability ratings for BNC sentences

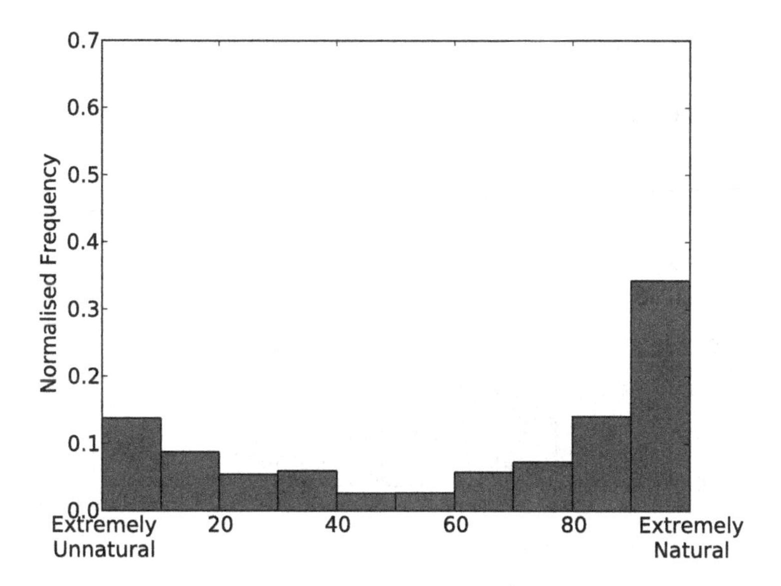

Figure 3.5 Slider individual acceptability ratings for BNC sentences.

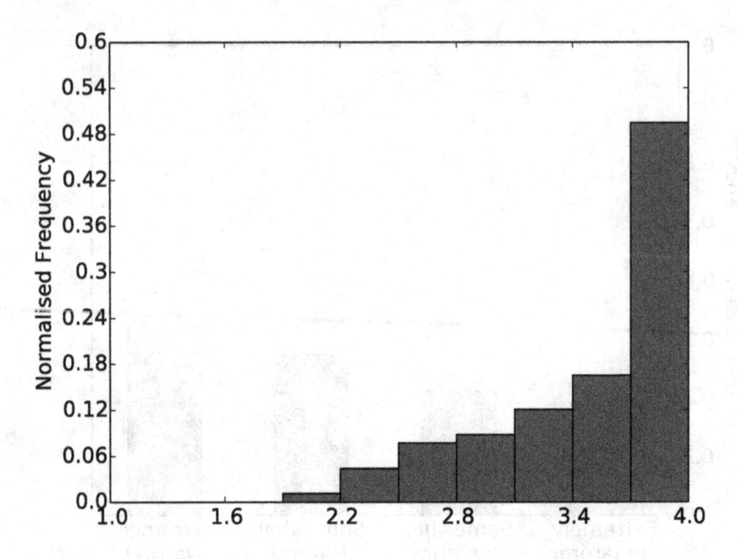

Figure 3.6 Mean acceptability ratings for good filtered Adger (2003) sentences.

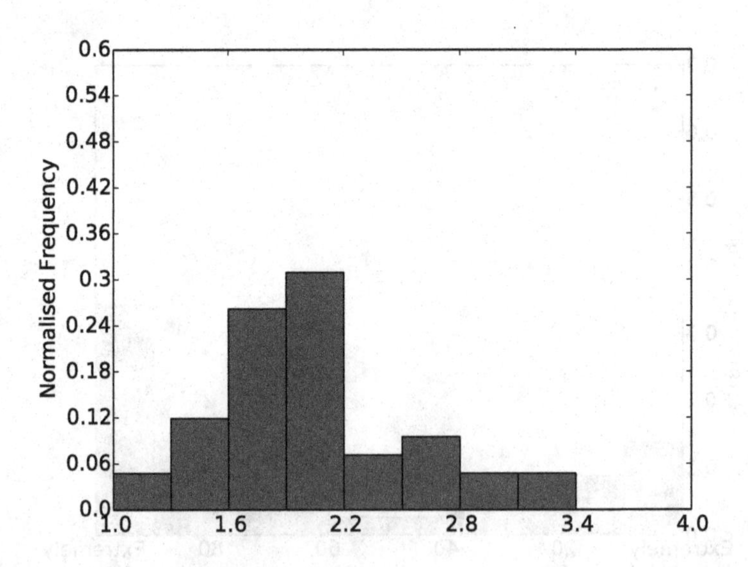

Figure 3.7 Mean acceptability ratings for starred filtered Adger (2003) sentences.

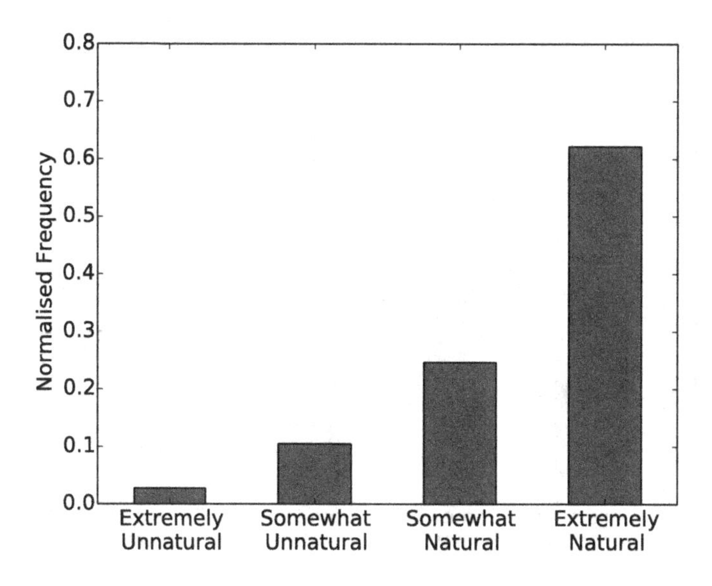

Figure 3.8 Individual acceptability ratings for good filtered Adger (2003) sentences.

3.2 PREDICTING ACCEPTABILITY WITH MACHINE LEARNING MODELS

Lau et al. (2015) and Lau, Clark, and Lappin (2017) train a series of unsupervised language models on the full BNC (100m words). They apply these models to a crowd-source (Amazon Mechanical Turk) annotated 2500 sentence test set obtained by round-trip machine translation, through four languages, from BNC original sentences. They filter the annotators for language fluency, and for reliability. They use a set of scoring functions to map the distributions that the models generate for the test set, into acceptability scores. They evaluate the accuracy of these scores through Pearson coefficient correlation with the mean speakers' judgements for the test set.

The primary classes of model that Lau et al. (2015) and Lau, Clark, and Lappin (2017) experiment with are

i. Lexical n-gram models (bigram, trigram, and 4-gram);

ii. A second-order Bayesian Hidden Markov Model (BHMM);

iii. A two-tier BHMM; and

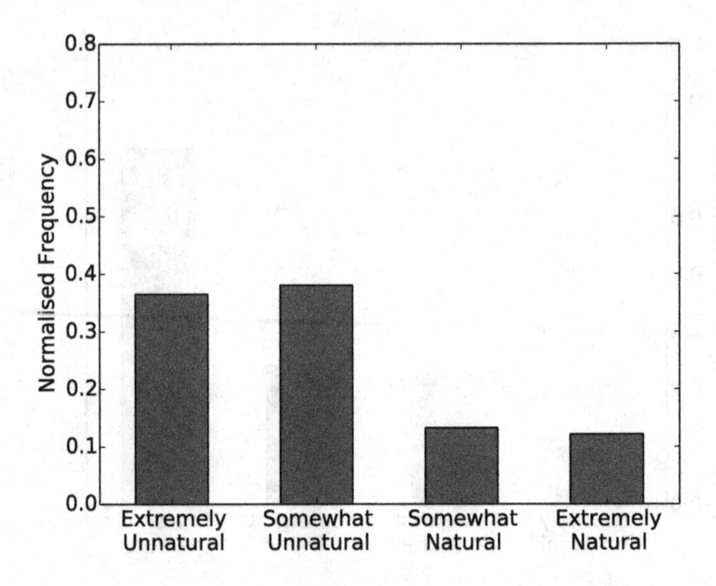

Figure 3.9 Individual acceptability ratings for starred filtered Adger (2003) sentences.

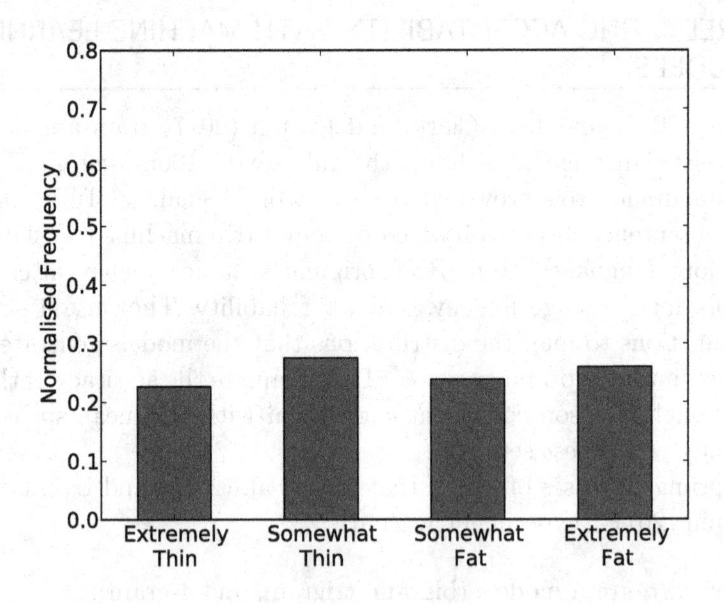

Figure 3.10 Individual body weight ratings for human figure drawings.

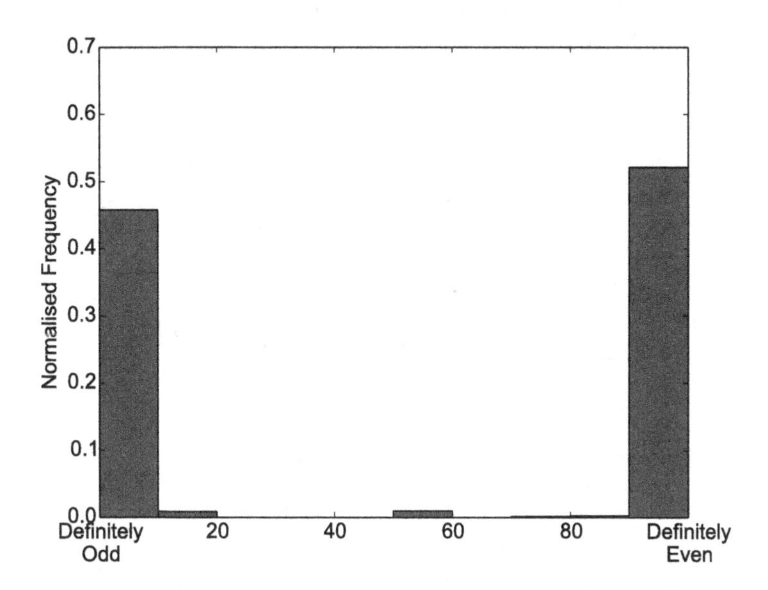

Figure 3.11 Individual even vs odd number ratings.

iv. A recurrent neural network language model (RNNLM).

For purposes of comparison, they also test the Stanford PCFG parser (a probabilistic Context-Free Grammar parser) (Klein & Manning, 2003a, 2003b).

In a simple n-gram LM, the conditional probability of a word is determined by the probabilities of the previous n words in the sequence in which it occurs (fewer, if the preceding sequence is shorter, or these words are out of vocabulary). This model is illustrated in Fig 3.12 with a trigram.

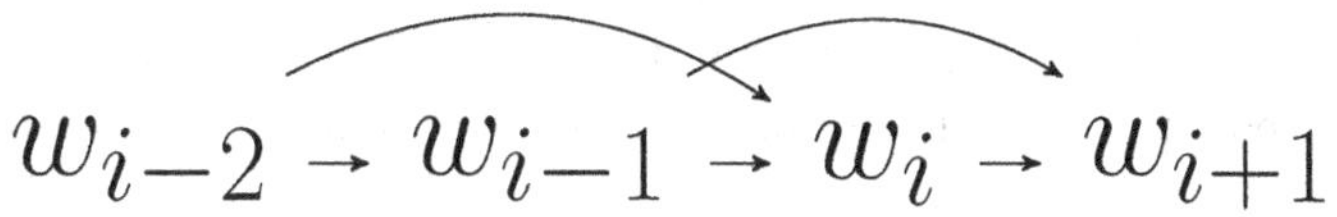

$$w_{i-2} \rightarrow w_{i-1} \rightarrow w_i \rightarrow w_{i+1}$$

Figure 3.12 Lexical Trigram

A Bayesian HMM first generates a (latent) word class, given its preceding word classes, and then it generates a word on the basis of the selected word class. A BHMM has two sets of multinomials: the state transition multinomials and the word emission multinomials. Fig 3.13 displays the structure of a BHMM.

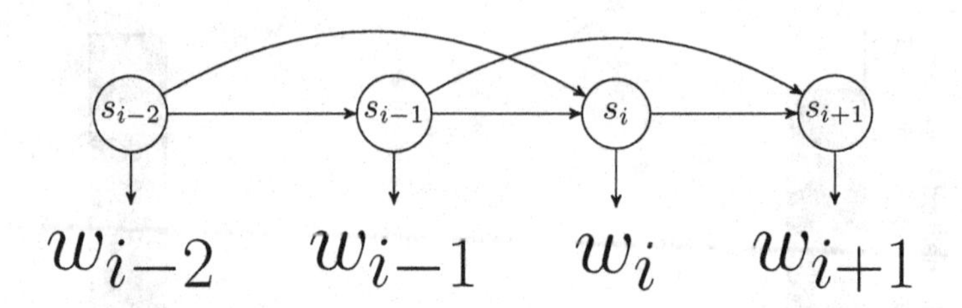

Figure 3.13 Bayesian HMM.

Lau et al. (2015) and Lau, Clark, and Lappin (2017)'s two-tier BHMM has an additional layer of latent variables on top of the word classes of a BHMM, where the variables in this second layer can be interpreted as phrase classes. The model conditions the prediction of word classes and words on these phrase classes. They use collapsed Gibbs sampling to infer phrase classes. They sample the tier-1 state s and tier-2 state t separately. The architecture of this model is given in Fig 3.14.[1]

Lau et al. (2015) and Lau, Clark, and Lappin (2017) use Mikolov et al. (2011)'s implementation of an RNNLM. Mikolov (2012) specifies an RNNLM that combines neural network learning with a Maximum Entropy (ME) component that learns direct connections among n-gram features.[2] They find that the efficacy of the ME component declines as the number of neural units increases, with neural network performance rendering ME insignificant after 500 units.

Lau et al. (2015) and Lau, Clark, and Lappin (2017) focus on unsupervised models because of what they can show us about the limits of human learning. To the extent that these models acquire linguistic

[1]Gibbs sampling is a Markov Chain Monte Carlo procedure for inferring a sequence of approximate observations from a multivariate probability distribution. In the two-tier HMM, it is used to infer phrase and word classes, which jointly condition the prediction of words.

[2]The Maximum Entropy principle specifies that for a class of probability distributions D over a set pf phenomena, in the absence of additional information about the elements of D, one selects the $d \in D$ with the highest entropy value as the default option.

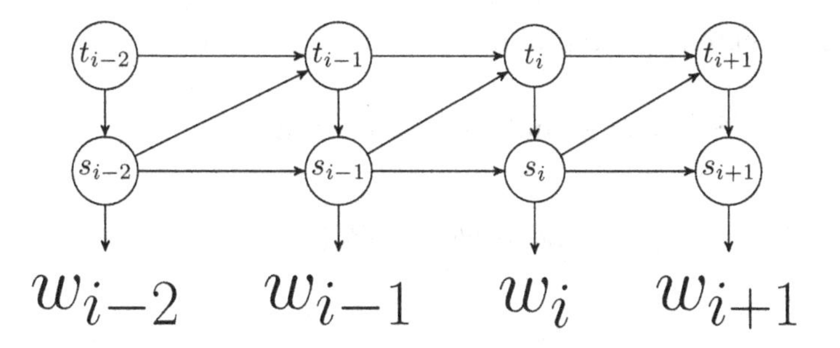

Figure 3.14 Two-Tier Bayesian HMM.

knowledge that allows them to accurately predict sentence acceptability, evaluated against a gold standard of mean human ratings, they indicate the sort of linguistic information that domain general machine learning systems can acquire from unlabelled text.

For purposes of comparison they also experiment with both the lexicalised and the unlexicalised Stanford PCFG parser (Klein & Manning, 2003a, 2003b), which is a supervised system. To compute the log probability of their test sentences, they use both the top-1 and the top-1K parses. The unlexicalised PCFG parser gives better performance, but they saw little difference between using the top-1 and the top-1K parses for computing log probability. It is important to note that because the Stanford PCFG is trained on a labelled corpus, through supervision, to generate a probability distribution over CFG parse trees for a test set, its performance on the current task does not provide a significant indication of its quality as a language model, or a parser. Lau et al. (2015) and Lau, Clark, and Lappin (2017) include it as a baseline for the unsupervised machine learning models that they test.

Lau et al. (2015) and Lau, Clark, and Lappin (2017) use scoring functions to map sentence logprob values into acceptability scores. These functions are designed to eliminate the effect of sentence length and lexical frequency. They measure the effect of the scoring functions by computing their correlations with both properties. They also compute the Pearson correlations of these properties with human acceptability judgements. They find that the correlation of human ratings and sentence length $= +0.13$, and of human ratings and minimum word

TABLE 3.1 Sentence Acceptability Scoring Functions

Scoring Function	**Equation**
LogProb	$\log P_m(\xi)$
Mean LP	$\frac{\log P_m(\xi)}{\|\xi\|}$
Norm LP (Div)	$-\frac{\log P_m(\xi)}{\log P_u(\xi)}$
SLOR	$\frac{\log P_m(\xi) - \log P_u(\xi)}{\|\xi\|}$

ξ = sentence;
$P_m(\xi)$ = the probability of the sentence given by the model;
$P_u(\xi)$ = is the unigram probability of the sentence = for
$w_i \in \xi \; (1 \leq i \leq k), \prod_{i=1}^{k} P(w_i)$;
SLOR is proposed by Pauls and Klein (2012)

frequency = +0.07. These low correlations support the view that human acceptability judgements are not determined by either factor.

The acceptability scores that the functions generate for a set of sentences in a test corpus do not sum to 1, and they are not probability values. These scores predict the relative acceptability of sentences, and they are based on the probability distribution of a language model. They specify a precise connection between probability and acceptability, and they give the predictions which Lau et al. (2015) and Lau, Clark, and Lappin (2017) use to evaluate the accuracy of their models against mean human ratings. The logprob and the three main functions that they use are defined in Table 3.1.

In addition to the sentence log probability scoring functions, Lau et al. (2015) and Lau, Clark, and Lappin (2017) also experiment with *individual word log probability*. For each test sentence they extract the five words that yield the lowest normalised log probability for the sentence (normalised using the word's log unigram probability). They take each of these values in turn as the score of the sentence. They denote these measures as *Word LP Min-N*. *Word LP Min-1* is the log probability given

TABLE 3.2 Pearson Correlations for BNC Trained and Tested Models

Measure	3-gram	BHMM	2T
LogProb	0.30	0.25	0.26
Mean LP	0.35	0.26	0.31
Norm LP (Div)	**0.42**	0.44	**0.50**
SLOR	0.41	**0.45**	**0.50**
Word LP Min-1	0.35	0.26	0.35
Word LP Min-2	0.41	0.38	0.43
Word LP Min-3	0.41	0.42	0.44
Word LP Min-4	0.40	0.43	0.43
Word LP Min-5	0.39	0.41	0.41

by the word with the lowest normalised log probability, *Word LP Min-2* the log probability given by the word with the second lowest normalised log probability, etc. This class of scoring functions seeks to identify the lexical locus of syntactic anomaly.

Table 3.2 gives the results of Lau et al. (2015) and Lau, Clark, and Lappin (2017)'s experiments with the trigram model, BHMM, and two-tier BHMM (2T), trained on the BNC, for predicting mean human judgements for their BNC test set, with each of the scoring functions. The highest scores are indicated in bold. The results for the BNC trained and tested RNNLM, with 600 neurons, are displayed in Table 3.3. Table 3.4 shows the correlations for the unlexicalised Stanford PCFG parser, tested on the BNC. The RNNLM, with *SLOR* (and *Norm LP (Div)*), outperforms the other models for this set of experiments. Pearson correlations over 0.4 are significant, and so the fact that a simple unsupervised RNN language model, trained on raw text, achieves 0.53 on this task is encouraging.

Lau et al. (2015) and Lau, Clark, and Lappin (2017) also test their BNC trained models on the AMT annotated, filtered Adger test set. As Tables 3.5 and 3.6 indicate, the trigram and two-tier BHMM do better than the RNNLM for this case. All three models score higher with *Word LP Min-N* functions than with any of the four global functions.

Training the models on English Wikipedia text improves the scores of the non-neural systems for the Adger test set, but not the RNNLM (Tables 3.7 and 3.8).The lower level of performance of all the models

TABLE 3.3 RNNLM BNC Trained and Tested

Measure	RNNLM
LogProb	0.32
Mean LP	0.39
Norm LP (Div)	**0.53**
SLOR	**0.53**
Word LP Min-1	0.38
Word LP Min-2	0.48
Word LP Min-3	0.50
Word LP Min-4	0.51
Word LP Min-5	0.50

TABLE 3.4 Stanford PCFG BNC Tested

Measure	Stanford PCFG (Unlexicalised)
LogProb	0.21
Mean LP	0.18
Norm LP (Div)	**0.26**
Norm LP (Sub)	0.22
SLOR	0.25

TABLE 3.5 3-gram, BHMM, 2T BNC Trained and Filtered Adger Tested

Measure	3-gram	BHMM	2T
LogProb	0.33	0.26	−0.21
Mean LP	0.27	0.11	0.18
Norm LP (Div)	0.36	0.30	0.17
SLOR	0.37	0.31	0.37
Word LP Min-1	**0.45**	0.10	0.32
Word LP Min-2	0.34	**0.34**	**0.40**
Word LP Min-3	0.34	0.41	**0.40**
Word LP Min-4	0.25	0.34	0.35
Word LP Min-5	0.33	0.26	0.30

TABLE 3.6 RNNLM BNC Trained and Filtered Adger Tested

Measure	RNNLM
LogProb	0.32
Mean LP	0.17
Norm LP (Div)	0.23
SLOR	0.23
Word LP Min-1	0.02
Word LP Min-2	0.27
Word LP Min-3	**0.38**
Word LP Min-4	0.28
Word LP Min-5	0.29

TABLE 3.7 English Wikipedia Trained and Filtered Adger Tested

Measure	3-gram	BHMM	2T
LogProb	0.34	0.33	0.35
Mean LP	0.28	0.23	0.26
Norm LP (Div)	0.36	**0.41**	0.41
SLOR	0.36	0.40	0.39
Word LP Min-1	**0.49**	0.25	0.46
Word LP Min-2	0.35	0.37	**0.49**
Word LP Min-3	0.34	**0.41**	0.35
Word LP Min-4	0.29	0.39	0.29
Word LP Min-5	0.32	**0.41**	0.29

on the Adger set may be due to the fact that most of its sentences are considerably shorter than both the BNC and the Wikipedia training sets. Moreover, the ill-formedness of the starred sentences seems to be local and lexically indicated, which may account for the fact that the models score higher with the *Word LP Min-N* functions in this experiment. We will see in the next chapter that bidirectional transformers improve considerably on these correlations for the Adger set, with a minimal global scoring function.

Lau et al. (2015) and Lau, Clark, and Lappin (2017) train and test their models on sentences from the English Wikipedia (ENWIKI). They follow the same protocol that they applied for their BNC experiment. They use AMT crowd-sourcing (filtered for language fluency and reliability) on round-trip Google translated sentences to obtain a test set

TABLE 3.8 RNNLM English Wikipedia Trained and Filtered Adger Tested

Measure	RNNLM
LogProb	0.35
Mean LP	0.23
Norm LP (Div)	0.27
SLOR	0.25
Word LP Min-1	0.04
Word LP Min-2	0.30
Word LP Min-3	**0.38**
Word LP Min-4	0.34
Word LP Min-5	0.28

TABLE 3.9 ENWIKI 3-Gram, BHMM, and 2T

Measure	3-gram	BHMM	2T
LogProb	0.36	0.32	0.35
Mean LP	0.36	0.28	0.35
Norm LP (Div)	0.41	0.44	0.49
SLOR	0.41	0.46	**0.50**
Word LP Min-1	0.38	0.36	0.37
Word LP Min-2	0.43	0.46	0.49
Word LP Min-3	0.43	0.47	**0.50**
Word LP Min-4	**0.44**	0.47	**0.50**
Word LP Min-5	0.43	**0.48**	0.49

of 2500 sentences. They train their models on 100m words of randomly selected English Wikipedia text. Tables 3.9 and 3.10 give the results of this experiment. The RNNLM yields the highest correlations, with both *SLOR* and several of the *Word LP Min-N* scoring functions.

They also train and test their models on Wikipedia texts in Spanish (ESWIKI), German (DEWIKI), and Russian (RUWIKI). They use the same protocol for round-trip machine translation and crowd-sourced AMT acceptability judgements to annotate test sentences in these languages, that they employed for the BNC and English Wikipedia experiments. The RNNLM, with *SLOR*, gives the best performance for these three Wikipedia corpora, achieving high Pearson scores (Table 3.11).

It is not reasonable to expect machine learning models to achieve a perfect correlation with mean judgements, given that individual human annotators could not do this. Lau et al. (2015) and Lau, Clark,

TABLE 3.10 ENWIKI RNNLM

Measure	RNNLM
LogProb	0.44
Mean LP	0.46
Norm LP (Div)	0.55
SLOR	0.57
Word LP Min-1	0.51
Word LP Min-2	0.60
Word LP Min-3	**0.62**
Word LP Min-4	0.60
Word LP Min-5	0.58

TABLE 3.11 RNNLM with SLOR for ESWIKI, DEWIKI, and RUWIKI

Corpora	RNNLM
ESWIKI	0.60
DEWIKI	0.69
RUWIKI	0.61

and Lappin (2017) estimate an arbitrary individual human annotator's performance relative to the set of mean judgements for a test set. They randomly select a single rating for each sentence, and they compute the Pearson correlation between these individual judgements and the mean ratings for the rest of the annotators (one vs the rest). They ran the experiment 50 times to reduce sample variation. The simulated individual human predictor specifies an upper bound on any model's expected performance on the sentence acceptability prediction task. Table 3.12 shows the Pearson correlations of one-vs the rest individual annotations to mean ratings for the test sets in Lau et al. (2015) and Lau, Clark, and Lappin (2017)'s experiments.

If we use estimated individual human performance to evaluate models, then Lau et al. (2015) and Lau, Clark, and Lappin (2017)'s best models (enriched with scoring functions) do quite well. 3-gram + *Word LP Min-1* and 2T + *Word LP Min-2*, trained on ENWIKI, both scored 0.49 for the Adger filtered test set, against an estimated individual human annotator correlation of 0.72. RNNLM + *Norm LP (Div)* and RNNLM + *SLOR*, trained and tested on the BNC, scored 0.53, against an estimated human correlation of 0.66. RNNLM + *Word LP Min-3*, trained and tested on an ENWIKI corpus, achieved 0.62, against

TABLE 3.12 Pearson Correlations for One vs the Rest to Mean Ratings

Test Domain	Corr to Mean
Adger Filtered	0.726
BNC	0.667
ENWIKI	0.741
ESWIKI	0.701
DEWIKI	0.773
RUWIKI	0.655

estimated human performance of 0.74. RNNLM + *SLOR* on DEWIKI is 0.69, against an estimated human performance of 0.77. RNNLM + *SLOR* on ESWIKI is 0.60, against an estimated human performance of 0.701. RNNLM + *SLOR* on RUWIKI is 0.61, against an estimated human performance of 0.65.

3.3 ADDING TAGS AND TREES

Ek, Bernardy, and Lappin (2019) test the effect of enriching the training data with syntactic and semantic annotations, on the performance of an LSTM LM in the sentence acceptability task. For a simple LSTM LM trained on raw text, the probability of a sentence is computed with

$$P_M(w_i) = P(w_i|(w_{i-1}), ..., (w_{i-n})).$$

An LSTM LM trained on text annotated with semantic or syntactic tags predicts the next word w_i in a sentence on the basis of the previous sequence of words and their tags. It uses

$$P_M(w_i) = P(w_i|(w_{i-1}, t_{i-1}), ..., (w_{i-n}, t_{i-n})).$$

The current tag (t_i) is not given when the model predicts the current word (w_i).

Ek et al. (2019) implement four variants of LSTM language models, each of which predicts the next word in a sequence, conditioned on

i. only the unannotated previous sequence of words;

ii. the previous sequence of words and their semantic role tags;

iii. the previous sequence of words and their syntactic dependency tags; and

iv. the previous sequence of words and their dependency tree depth indicators.

Ek et al. (2019)'s LSTMs are unidirectional, with one level of 600 units. The models are trained on a vocabulary of 100,000 words. They use word embeddings of 300 dimensions, and 30 dimensions for tags. They apply a drop out of 0.4 after the LSTM layer. Training runs for 10 epochs. Ek et al. (2019) train their LMs on a randomly selected subset of the CoNLL 2017 dataset (Nivre et al., 2017). The corpus is annotated with dependency parse trees. They remove sentences whose dependency root is not a verb to eliminate non-sentences. They also delete sentences longer than 30 words. The remaining corpus contains 87M tokens and 5.3M sentences.

They use Lau et al. (2015) and Lau, Clark, and Lappin (2017)'s AMT annotated test set of 2500 BNC sentences as their test suite. This set contains 500 original English sentences and 2000 sentences translated through Norwegian, Spanish, Chinese, or Japanese back to English. Annotation is on a scale of 1 to 4. On average, each sentence is rated by 14 annotators.

Ek et al. (2019) automatically tag the training corpus and the test set with semantic role markers obtained from Abzianidze et al. (2017)'s Parallel Meaning Bank. These roles provide a fine-grained set of semantic types for expressions of major lexical categories. So, for example, the semantic roles of *he* and *his* are distinguished in

He	*took*	*his*	*book*	.
`PRO`	`EPS`	`HAS`	`CON`	`NIL`

Nivre et al. (2016, 2017)'s Universal Dependency grammar parses sentences with trees that specify dependency relations among its constituents, as in Fig 3.15.

Ek et al. (2019) use D. Chen and Manning (2014)'s Stanford Dependency Parser to generate syntactic tags for the training and test sets. They follow Gómez-Rodríguez and Vilares (2018) in encoding the tree depth of a word n by computing the number of common ancestors in the

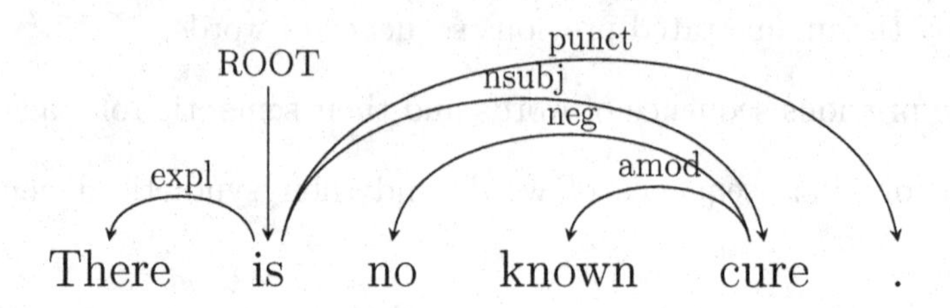

Figure 3.15 Labelled Universal Dependency Parse Tree.

tree between word n and word $n + 1$. Dependency tags combined with depth indicators provide a linearised encoding of a dependency tree, as in Fig 3.16.

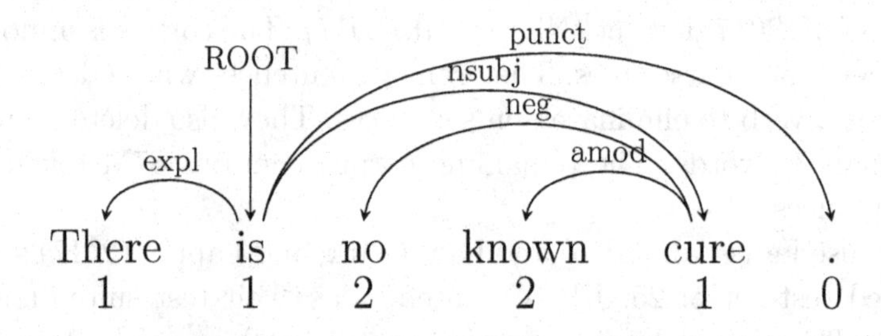

Figure 3.16 Linearised Encoded Dependency Parse Tree.

They use *SLOR* as the acceptability scoring function for their models. Instead of the simple Pearson correlation coefficient, they apply a weighted Pearson correlation to assess their model's performance in predicting acceptability, relative to human ratings. The weighted Pearson metric takes account of the fact that the model's score is compared to the mean rating value of a sentence, derived from several annotators, by giving greater weight to ratings with lower standard deviation.

The matrix in Table 3.13 shows the weighted Pearson correlations for Ek et al. (2019)'s models. LSTM is the model trained on unannotated Wikipedia text. +SYN, + SEM, +DEPTH are the LSTM with a training corpus annotated with syntactic dependency tags, semantic role tags, and tree depth encoding, respectively. The +Syn+Depth model is trained on text with linearised tree encodings. $M*$ indicates a version of the model M with a training corpus in which the tags are randomly permuted.

TABLE 3.13 Ek et al. (2019) Model Performance

	HUMAN	LSTM	+SYN	+SYN*	+SEM	+SEM*	+DEPTH
HUMAN	1.00						
LSTM	**0.58**	1.00					
+SYN	0.55	0.96	1.00				
+SYN*	0.39	0.76	0.75	1.00			
+SEM	0.54	0.81	0.78	0.61	1.00		
+SEM*	0.52	0.81	0.78	0.63	0.96	1.00	
+DEPTH	0.56	0.97	0.97	0.74	0.79	0.79	1.00
+DEPTH*	0.46	0.87	0.85	0.73	0.72	0.72	0.86
+SYN+DEPTH	0.54						

The simple LSTM LM outperforms all models trained on text with syntactic and semantic tags, achieving 0.58 correlation to mean human judgements (comparable to Lau et al. (2015) and Lau, Clark, and Lappin (2017)'s results for RNNLM with *SLOR*). The depth indicator model does best of all annotated models (0.56), the syntactic tag model is just below it (0.55), while the semantic role model scores lowest (0.54). Shuffling the tags causes a drop of 0.16 in correlation for syntactic tags, 0.1 for tree depth, but only 0.02, for semantic tags, indicating that syntactic tags and depth markers contribute more information to their respective models' predictions. The combined syntactic tag + tree depth marker model (0.54) performs below each of its component models.

Ek et al. (2019) measure the perplexity of a model by its cross entropy training loss.[3] Table 3.14 shows the perplexity (cross entropy loss) values for each model.

TABLE 3.14 Model Perplexity

MODEL	LOSS	ACCURACY
LSTM	5.04	0.24
+SYN	**4.79**	0.26
+SEM	5.23	0.21
+DEPTH	4.88	0.27

There does not appear to be a direct correlation between an LSTM's quality as a language model, as indicated by its perplexity, and its per-

[3]As we noted in Chapter 2, fn 3, perplexity is the logarithm (exponentiation) of entropy.

formance on the sentence acceptability task. Syntactic tags and depth indicators decrease perplexity, but semantic markers increase it. The simple non-annotated LSTM outperforms all of them. It might be that *SLOR* masks the underlying perplexity of these models.

Warstadt, Singh, and Bowman (2019) discuss several pre-trained transformer models applied to classifying sentences in their Corpus of Linguistic Acceptability (CoLA) as acceptable or not. These models exhibit levels of accuracy that vary widely relative to the types of syntactic and morphological patterns that appear in CoLA. This is a very different sort of test set from the one that Lau et al. (2015), Lau, Clark, and Lappin (2017), and Ek et al. (2019) use in their experiments. It is drawn from linguists' examples intended to illustrate particular sorts of syntactic construction, and annotated with linguists' binary judgments. By contrast, the BNC test set consists of naturally occurring text, where a wide range of infelicities are introduced into many of the sentences through round-trip machine translation, and it is annotated through AMT crowd-sourcing with gradient acceptability judgments. I return to Warstadt et al. (2019)'s work in Chapter 5.

3.4 SUMMARY AND CONCLUSIONS

In this chapter, we have considered the connection between grammaticality, acceptability, and probability. Grammaticality is a theoretical property not directly accessible to observation. By contrast, human acceptability ratings are primary data that we can measure. Linguists have generally used these judgements to support claims about grammaticality. LCL provide extensive evidence that gradience is pervasive in individual, as well as aggregate, acceptability judgements. An adequate model of linguistic representation must accommodate the gradience, and accurately predict the observed distribution of ratings for sentences.

Lau et al. (2015) and Lau, Clark, and Lappin (2017) specify a set of functions that map a sentence's probability (logprob) values into acceptability scores. These neutralise the effects of sentence length and word frequency on relative acceptability predictions. They specify a precise connection between a sentence's probability within the distribution that a model generates for a corpus, and its predicted acceptability.

We looked at a series of experiments in which the acceptability predictions of unsupervised language models are evaluated through Pearson correlations with mean human ratings obtained through AMT crowd-source annotation. Infelicities are introduced into some of the test sets through round-trip machine translation. A simple unsupervised RNN LM performs surprisingly well on the sentence acceptability task for BNC and Wikipedia test sets. It did less well on a filtered set of linguists' examples, but this may well be due to the short length of these sentences, and the fact that they are significantly different in kind and structure from the natural text on which the models are trained.

Ek et al. (2019) observe that an unsupervised LSTM LM trained on a Wikipedia corpus and tested, out of domain, on Lau et al. (2015) and Lau, Clark, and Lappin (2017)'s BNC test set, approximated Lau et al. (2015) and Lau, Clark, and Lappin (2017)'s RNN LM results. This indicates the robustness of the LSTM. They also found that enhancing the input of the LSTM with syntactic or semantic markers, or full tree structures, undermines its predictive power for the sentence acceptability task. These results provide additional support for Bernardy and Lappin (2017)'s finding that using abstract syntactic markers to highlight structural relations may degrade an LSTM's performance on certain tasks. More generally, it is consistent with the results that we discussed in the previous chapter, that suggest that introducing structural bias into the training data or the architecture of an LSTM does not significantly improve its performance across a range of NLP tasks requiring syntactic knowledge. We will return to this issue in Chapter 5.

CHAPTER 4

Predicting Human Acceptability Judgements in Context

4.1 ACCEPTABILITY JUDGEMENTS IN CONTEXT

Linguists and cognitive scientists have frequently noted the fact that the interpretation of a sentence can be significantly influenced by the contexts, linguistic, and extra-linguistic, in which it appears. Context is, then, an important factor in determining the relative acceptability of a sentence.

Lau, Clark, and Lappin (2014, 2015, 2017) and Lau, Clark, and Lappin (2017) (LCL) and Ek, Bernardy, and Lappin (2019) test speakers' acceptability judgements for sentences presented outside of any context beyond the HIT (Human Intelligence Task) in which they appear. The sentences in each HIT are randomly selected. Their models predict acceptability without reference to context. Document context is not explicitly represented in any of the models in either training or testing. Bernardy, Lappin, and Lau (2018) construct two datasets of sentences annotated with acceptability ratings, one judged with, and the other without document context. They extracted 100 random articles from the English Wikipedia, and they sampled a sentence from each article.

They tried LCL's method of using Google Translate for round-trip MT to generate a set of sentences with varying degrees of acceptability. Google Translate has improved to the point that a pilot study indicated that human annotators rated most round-trip translated sentences as highly as the English originals. This indicates a substantial improvement in the quality of Google Translate for the language pairs tested for round-trip translation, during the period between the LCL experiments and Bernardy et al. (2018)'s pilot studies. This improvement may well have been driven by the move to powerful neural language models supporting encoder-decoder MT, with the addition of attention.

As an alternative to Google Translate, Bernardy et al. (2018) used the more traditional statistical phrase based Moses MT system (Koehn et al., 2007). They applied the pre-trained Moses models for round-trip MT into Czech, Spanish, German, and French, and then back to English. This provided a distribution of acceptability judgements over sentences comparable to those which LCL obtained in their experiments. Following LCL's protocol, Bernardy et al. (2018) used HITS with a four category acceptability rating for crowd source annotation of both out-of-context and in-context datasets.

The target sentence was highlighted in boldface, with one preceding and one succeeding sentence included as additional context. Annotators had the option of revealing the full document context by clicking on the preceding and succeeding sentences. As in the out-of-context test set, sentences were presented in HITS of five, one from the original English set, and four from the round-trip translations. Each HIT contained one sentence per target language, with no sentence type appearing more than once in a HIT.

The examples in 4.1 show the mean out of context and in context ratings for a sentence from their test sets, under round-trip translation with Moses, through the four languages. The original English sentence appears first in the list.

4.1(a) **English original**: david acker, harry's son, became the president of sleepy's in 2001.
$\mathtt{human}^{-context}$ 3.47 $\mathtt{human}^{+context}$ 3.38

(b) **Czech**: david acker harry' with son has become president of the sleepy' with in 2001.
$\mathtt{human}^{-context}$ 1.75 $\mathtt{human}^{+context}$ 2.08

(c) **German**: david field, harry' the son was the president of" in 2001.
$\mathtt{human}^{-context}$ 1.63 $\mathtt{human}^{+context}$ 3.00

(d) **Spanish**: david acker, harry' his son, became president of the sleeping' in 2001.
$\mathtt{human}^{-context}$ 2.19 $\mathtt{human}^{+context}$ 2.62

(e) **French**: david acker, harry' son, the president of the sleepy' in 2001.
human$^{-context}$ 1.47 **human**$^{+context}$ 2.46

Fig 4.1 shows Bernardy et al. (2018)'s regression diagram for AMT annotated sentences presented in, and out of, document context.[1] The x-axis diagonal is the out-of-context line, and the y-axis is the in-context line.

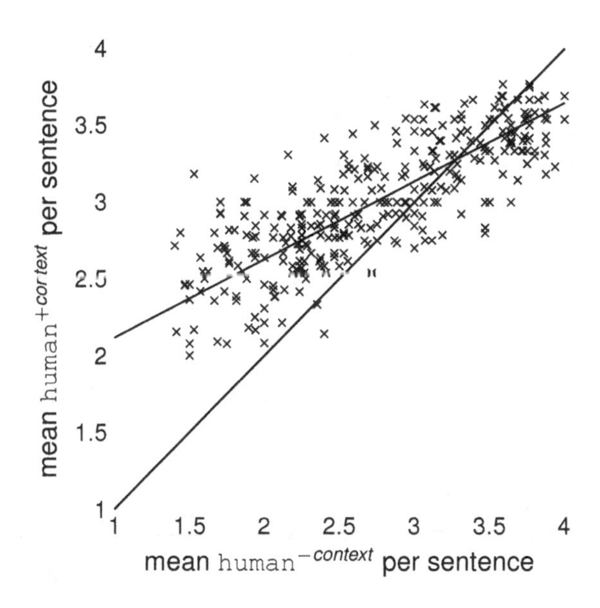

Figure 4.1 Bernardy et al. (2018)'s human ratings for in- and out-of-context sentences.

Bernardy et al. (2018) found a strong Pearson's r correlation of 0.80 between mean out-of-context and in-context judgements. The average difference between human$^{-context}$ and human$^{+context}$ is represented by the distance between the linear regression and the full diagonal in the graph. These lines cross at human$^{+context}$ = human$^{-context}$ = 3.28, the point where context no longer boosts acceptability. Adding context

[1]Statistical regression is used to determine the relation between a dependent variable to one or more independent variables. The linear regression graph in Fig 4.1 indicates the correlation between mean human ratings of sentences in context with those for the same sentences annotated out of context.

generally improves acceptability but the pattern reverses as acceptability approaches maximal mean rating values. This "compresses" the distribution of (mean) ratings, pushing the extremes to the middle. The net effect of this compression lowers correlation, as the good and bad sentences for the in-context test set are not as clearly separable as they are in the out-of-context test set.

Bizzoni and Lappin (2019) test the effect of context on gradient judgements of paraphrase for a metaphorical sentence. They solicit AMT crowd-source ratings for pairs containing a metaphorical sentence, and a candidate for a literal paraphrase of that sentence. In one test set 200 pairs are rated on a four category scale of paraphrase appropriateness, independently of context. In the second test set the same pairs are judged within a context of a preceding and a following sentence.

The two sentence triples in 4.2 are an example pair from Bizzoni and Lappin (2019)'s second test set. 4.2(a) is a metaphorical sentence (in boldface) embedded in a context, and 4.2(b) is a candidate paraphrase of this sentence, in the same context.

4.2(a) They had arrived in the capital city. **The crowd was a roaring river**. It was glorious.

(b) They had arrived in the capital city. **The crowd was huge and noisy**. It was glorious.

Bizzoni and Lappin (2019) observe the same compression effect with in-context paraphrase judgements that Bernardy et al. (2018) obtain for in-context acceptability ratings. Fig 4.2 gives Bizzoni and Lappin (2019)'s regression for in- and out-of-context AMT paraphrase judgements.

Bernardy et al. (2018) experiment with two DNN language models to predict the human sentence ratings for each of their test sets. `lstm` is a standard LSTM language model, trained over a corpus to predict word sequences. `tdlm` (Lau, Baldwin, & Cohn, 2017) is a topic driven neural LM. The topic model component of `tdlm` produces topics by processing documents through a convolutional layer, and aligning them with trainable topic embeddings. The language model component of `tdlm` incorporates context by combining its topic vector with its LSTM's hidden state, to generate the probability distribution for the next word.

Both LMs can use the document context as a prefix input to the sentence at test time. This yields 4 variant LMs at test time:

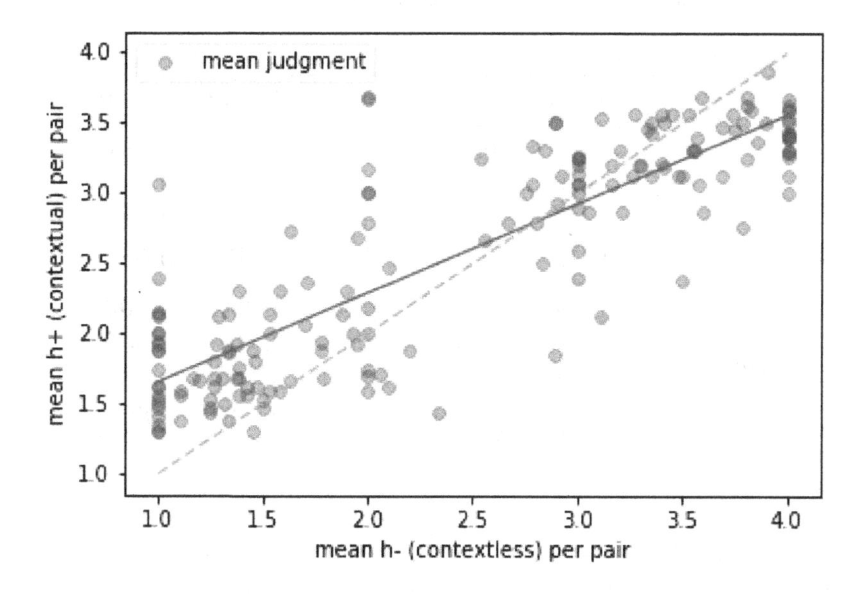

Figure 4.2 Bizzoni and Lappin (2019)'s regression graph for paraphrase ratings.

- lstm^{-c} and tdlm^{-c}, which use only sentences from a test set as input; and
- lstm^{+c} and tdlm^{+c}, which use sentence and context at test time.

To map sentence probability to acceptability Bernardy et al. (2018) use Lau et al. (2015) and Lau, Clark, and Lappin (2017)'s three scoring functions, defined in Chapter 3, Table 3.1. They train tdlm and lstm on a sample of 100K English Wikipedia articles, which has no overlap with the 100 documents used for test set annotation. The training data has approximately 40M tokens and a vocabulary size of 66K. To assess the performance of the acceptability measures, they compute Pearson's r against mean human ratings. They also experimented with Spearman's rank correlation but found similar trends, and so they present only the Pearson results. Tabel 4.1 shows the performance of their models.

lstm^{-c} against human$^{-context}$ with *SLOR* achieves 0.584, slightly surpassing the performance of RNNLM with *SLOR* in the original Lau et al. (2015) and Lau, Clark, and Lappin (2017) experiment (0.570). Across all models (lstm and tdlm) and human ratings (human$^{-context}$ and

TABLE 4.1 Bernardy et al. (2018) Model Performance

Rtg	Model	*LP*	*Mean*	*NrmD*	*SLOR*
human$^{-context}$	lstm^{-c}	0.151	0.487	**0.586**	0.584
	lstm^{+c}	0.161	0.529	0.618	**0.633**
	tdlm^{-c}	0.147	0.515	0.634	**0.640**
	tdlm^{+c}	0.165	0.541	0.645	**0.653**
human$^{+context}$	lstm^{-c}	0.153	0.421	0.494	**0.503**
	lstm^{+c}	0.168	0.459	0.522	**0.546**
	tdlm^{-c}	0.153	0.450	0.541	**0.557**
	tdlm^{+c}	0.169	0.473	0.552	**0.568**

human$^{+context}$), using context at test time improves model performance. Taking context into account helps in modelling acceptability, regardless of whether it is tested against judgements made with (human$^{+context}$) or without (human$^{-context}$) context. tdlm consistently outperforms lstm over both types of human ratings and test input variants. Context helps in the modelling of acceptability, whether it is incorporated during training (lstm vs. tdlm) or at test time (lstm^{-c}/tdlm^{-c} vs. lstm^{+c}/tdlm^{+c}).

The *SLOR* correlation of lstm^{+c}/tdlm^{+c} vs. human$^{+context}$ (0.546/568) is lower than that of lstm^{-c}/tdlm^{-c} vs. human$^{-context}$ (0.584/0.640). human$^{+context}$ ratings are more difficult to predict than human$^{-context}$. This raises the question as to why context reduces the spread between ratings. One possible explanation is that annotators focus more on discourse coherence when rating sentences in a document context. The issue of discourse coherence does not arise in human$^{-context}$ judgements. If this factor is, in fact, significant in annotation, then syntactic infelicities introduced by round-trip MT may play less of a role in rating for the human$^{+context}$ set.

A second explanation is that context imposes additional cognitive load, which reduces the speaker's/hearer's resources for identifying syntactic and semantic anomaly in an individual sentence.[2] If the discourse coherence account is correct, then we would expect the compression

[2]See Sweller (1988), Ito, Corley, and Pickering (2018), Causse, Peysakhovich, and Fabre (2016), and Park et al. (2013) for studies of the effect of cognitive load on human performance of primary tasks, both linguistic, and perceptual.

effect to be prominent with coherent contexts, but not with random contexts, which prevent integration of the sentence into a discourse unit. By contrast, the general cognitive load explanation predicts that the compression effect should be observable for both types of context, as each of them causes distraction through use of additional processing resources.

4.2 TWO SETS OF EXPERIMENTS

Following Bernardy et al. (2018)'s protocol, Lau, Armendariz, Lappin, Purver, and Shu (2020) generate a test set of 250 sentences from 50 English Wikipedia sentences, through round-trip MT, with Moses. They split the test set into 25 HITs of 10 sentences. Each HIT contains 2 original English sentences and 8 translated sentences, which are different from each other and not derived from either of the originals. Lau et al. (2020) use AMT crowd-sourcing to annotate the sentences for naturalness on a four point scale, for three types of context.

They present the sentences in each HIT in null, real, and, random contexts, respectively. Each context experiment was performed by a disjoint group of annotators. The real contexts consist of the three sentences that immediately precede a sentence in its document. The random contexts are consecutive sequences of three sentences taken from other documents.

In the context experiments Lau et al. (2020) first show the context paragraph, and they ask users to select the most appropriate description of its topic from a list of four candidate topics. Each candidate topic is represented by three words generated with a topic model. After performing this task the annotator is shown the sentence to be rated for acceptability. This experimental set up insures that annotators read the context sentences before assessing the sentences of the HIT. Annotators are filtered for reliability in topic identification and in sentence rating, and for language fluency. A high percentage of them satisfied the filtering criteria. Fig 4.3 shows a real context task from Lau et al. (2020), and Fig 4.4 gives the sentence rated for this context.

Lau et al. (2020) follow Hill et al. (2015)'s procedure for modulating the mean annotation results to eliminate the effect of outlier judgements. They calculate the average rating for each user, and the overall average by taking the mean of all average ratings. They decrease (increase) the ratings of a user by 1.0 if their average rating is greater (smaller) than the overall average by 1.0. To reduce the impact of outliers, for each

Read the following text and choose the three word set that you think comes closest to describe the topic:

```
einstein's views about religious belief have been
collected from interviews and original writings.
he called himself an agnostic, while disassociating
himself from the label atheist.
he said he believed in the "pantheistic" god of
baruch spinoza, but not in a personal god. a belief
he criticized.
```

Choose one of the following answers

- acid, carbon, metal
- church, catholic, christian
- population, age, women
- computer, code, systems

Figure 4.3 Real Context Task from Lau et al. (2020)

```
einstein once wrote: "i do not believe in a personal god
and i have never denied this but expressed it clearly".
```

bad **not very good** **mostly good** **good**

Figure 4.4 Example sentence from Lau et al. (2020).

sentence they remove ratings that are more than 2 standard deviations away from the mean. The pair wise regression graphs for the annotation sets are shown in Figs 4.5–4.7.

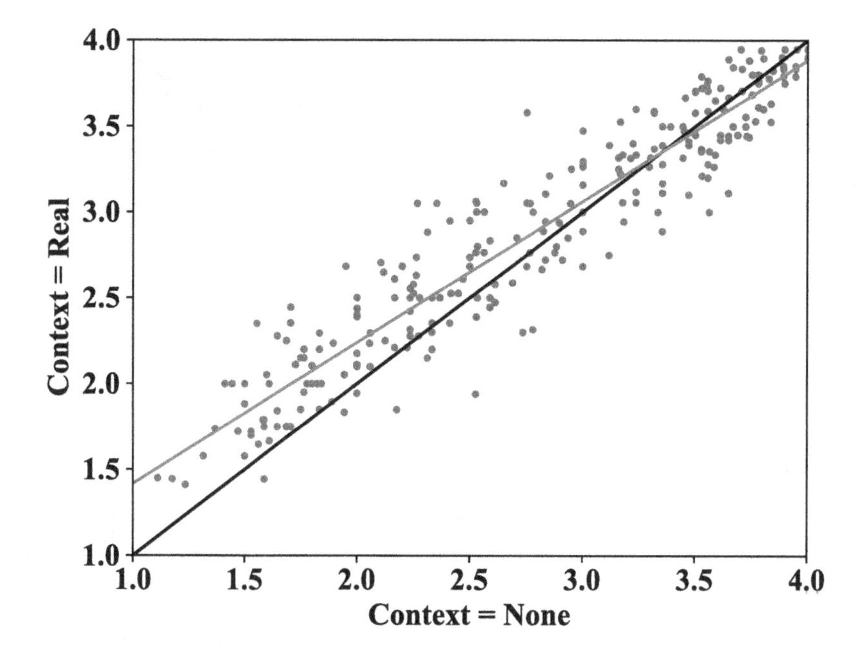

Figure 4.5 Lau et al. (2020)'s real context vs no context ratings.

The compression effect appears to be present in both real vs out-of-context, and random vs out-of-context ratings (Figs 4.5 and 4.6), although it specifies a different cross-over point in each case. However, it has been observed that when linear regression is applied to two sets of very noisy data, with outliers in both variables, it will yield a distributional pattern that resembles the compression effect exhibited in these regression graphs.[3]

To determine whether this effect is an artefact of regression to the mean for the test set annotations, Jey Han Lau applied total least squares errors-in-variables regression to the data. This is a least squares regression modelling procedure that allows for errors in both the variables between which a correlation is tested, rather than assuming that one of the variables is error free, as does linear (non-total least squares) regression. Lau also used this procedure with a swap of the dependent and independent variables, which involves permuting the x and y axes. The resulting graphs closely resemble our original linear regression patterns

[3]We are grateful to Roger Levy for pointing this out to us during discussion of Lau et al. (2020) at ACL 2020.

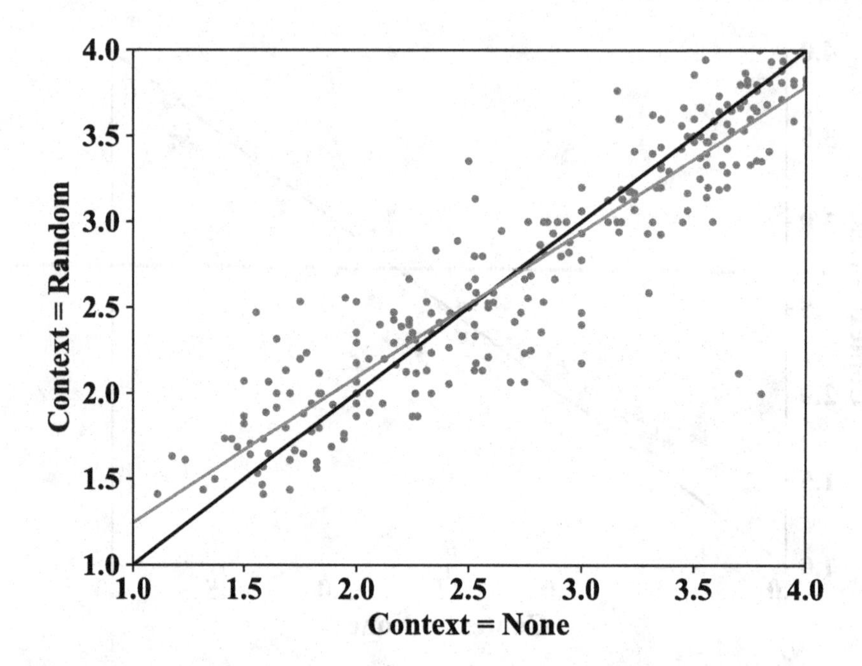

Figure 4.6 Lau et al. (2020)'s random context vs no context ratings.

for the corresponding variable pairs, with mirror image graphs for their permuted variants. This result strongly indicates that the compression effect is a real property of the data, rather than an epiphenomenon caused by regression to the mean. Figs 4.8–4.11 display the total least squares regression graphs for real vs no context ratings, random vs no context ratings, and their respective permutations.

4.3 THE COMPRESSION EFFECT AND DISCOURSE COHERENCE

The compression effect appears in both the h^+ (real context) vs. $h^\varnothing$ (null context), and the h^- (random context) vs. $h^\varnothing$ cases, for linear and for TLS regression. In addition, the h^+ vs. $h^\varnothing$ linear and TLS regression diagrams exhibit a raising effect in real contexts, which pushes the cross over point towards the upper end of the scale. In the h^- vs. h^+ figure the regression line is parallel to and below the diagonal, indicating a consistent decrease in acceptability ratings from h^+ to h^-. These effects suggest that the cognitive load of processing contexts produces

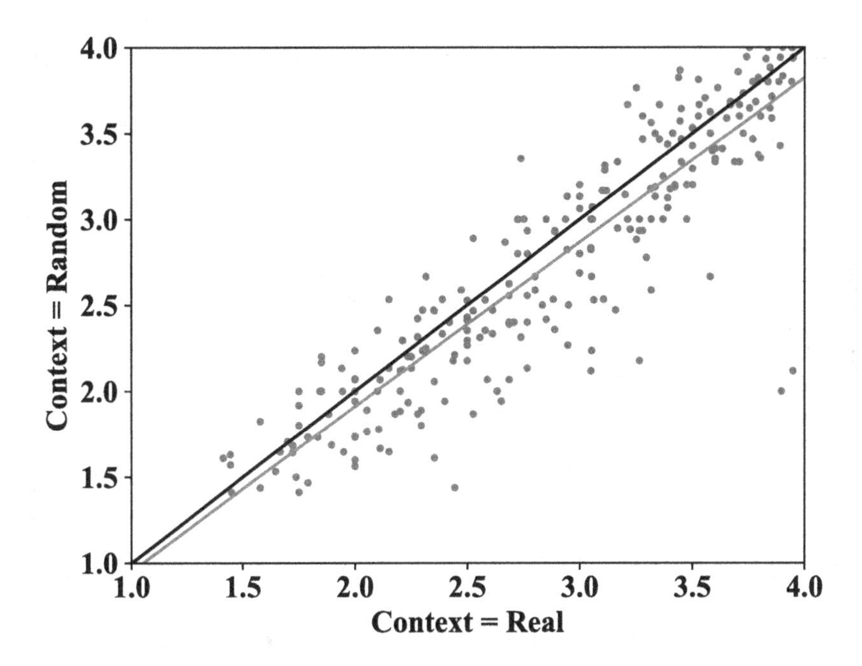

Figure 4.7 Lau et al. (2020)'s random context vs real context ratings.

compression in both $\mathtt{h}^+$ and $\mathtt{h}^-$, while discourse coherence operates only in $\mathtt{h}^+$ to generate a raising of acceptability ratings.

The mean ratings in all three test sets correlate strongly with each other, with Pearson's r for $\mathtt{h}^+$ vs. $\mathtt{h}^\varnothing = 0.945$, $\mathtt{h}^-$ vs. $\mathtt{h}^\varnothing = 0.917$, and $\mathtt{h}^-$ vs. $\mathtt{h}^+ = 0.901$. Lau et al. (2020) use the non-parametric Wilcoxon signed-rank test (one-tailed) to compare the difference between $\mathtt{h}^+$ and $\mathtt{h}^-$. The test gives a p-value of 2.4×10^{-8}, indicating that the discourse coherence effect is significant.

Lau et al. (2020) also use the Wilcoxon test to compare the regression lines for $\mathtt{h}^+$ vs. $\mathtt{h}^\varnothing$, and $\mathtt{h}^-$ vs. $\mathtt{h}^\varnothing$, to see if their offsets (constants) and slopes (coefficients) are statistically different. The p-value for the offset is 2.1×10^{-2}, confirming that there is a significant discourse coherence effect. The p-value for the slope, however, is 3.9×10^{-1}, suggesting that cognitive load compresses the ratings in a consistent way for both $\mathtt{h}^+$ and $\mathtt{h}^-$, relative to $\mathtt{h}^\varnothing$.

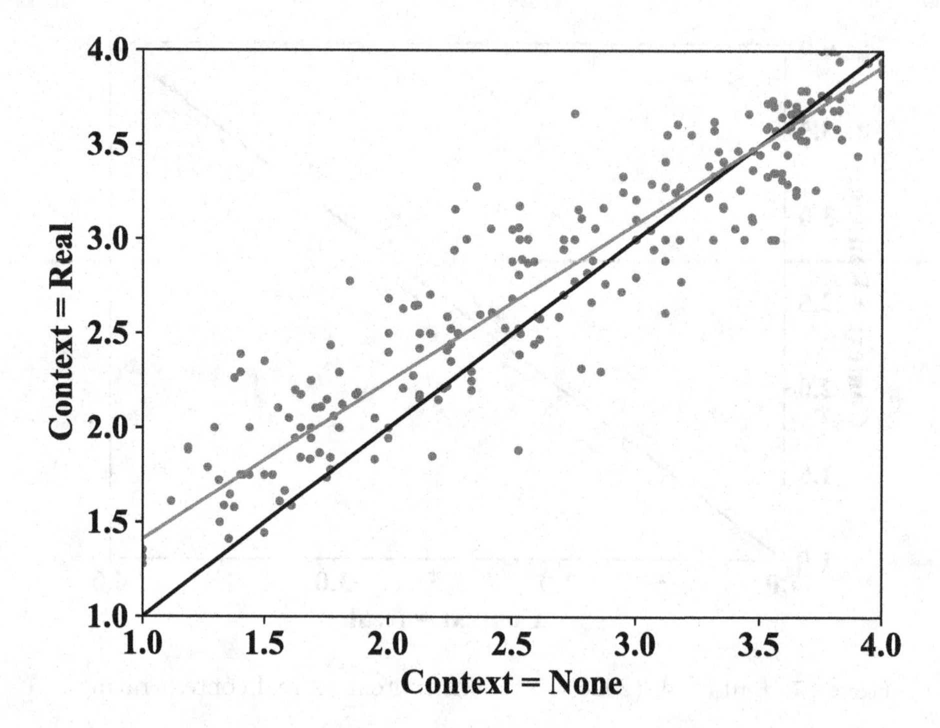

Figure 4.8 Lau's TLS regression for real context vs no context ratings.

4.4 PREDICTING ACCEPTABILITY WITH DIFFERENT DNN MODELS

In addition to `lstm` and `tdlm` Lau et al. (2020) experiment with three transformer language models. These are `gpt2`, `bert`, and `xlnet` (Yang et al., 2019). These models are equipped with large pre-trained lexical embeddings, and they apply multiple self-attention heads to all input words. As we noted in previous chapters, `bert` processes input strings without regard to sequence, in a massively parallel way, which permits it to efficiently identify large numbers of co-occurrence dependency patterns among the words of a string.

`lstm` and `gpt2` are unidirectional, and so they can be used to compute the probability of a sentence left to right, according to the formula

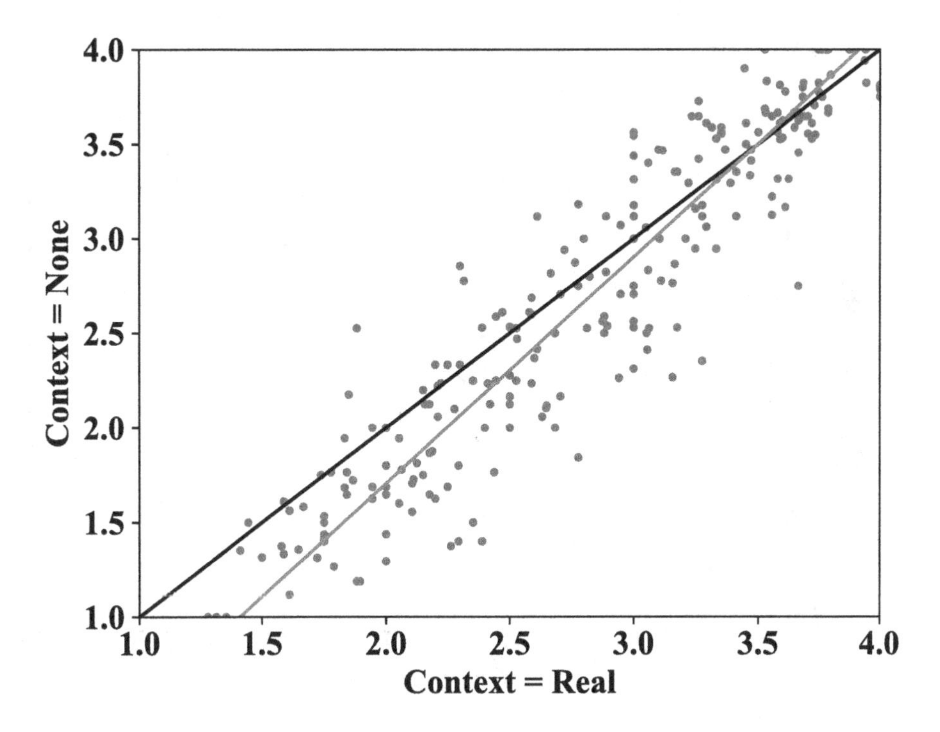

Figure 4.9 Lau's TLS regression for no context vs real context ratings.

$$P(s) = \prod_{i=0}^{|s|} P(w_i | w_{<i}).$$

`bert` is bidirectional, and it predicts words for both their left and right contexts. It requires the formula

$$P(s) = \prod_{i=0}^{|s|} P(w_i | w_{<i}, w_{>i}).$$

This equation does not yield true probabilities, as its values do not to sum to 1 (normalising these values to genuine probabilities is intractable). Instead these values provide confidence scores of likelihood. `xlnet` can be applied either unidirectionally or bidirectionally. Table

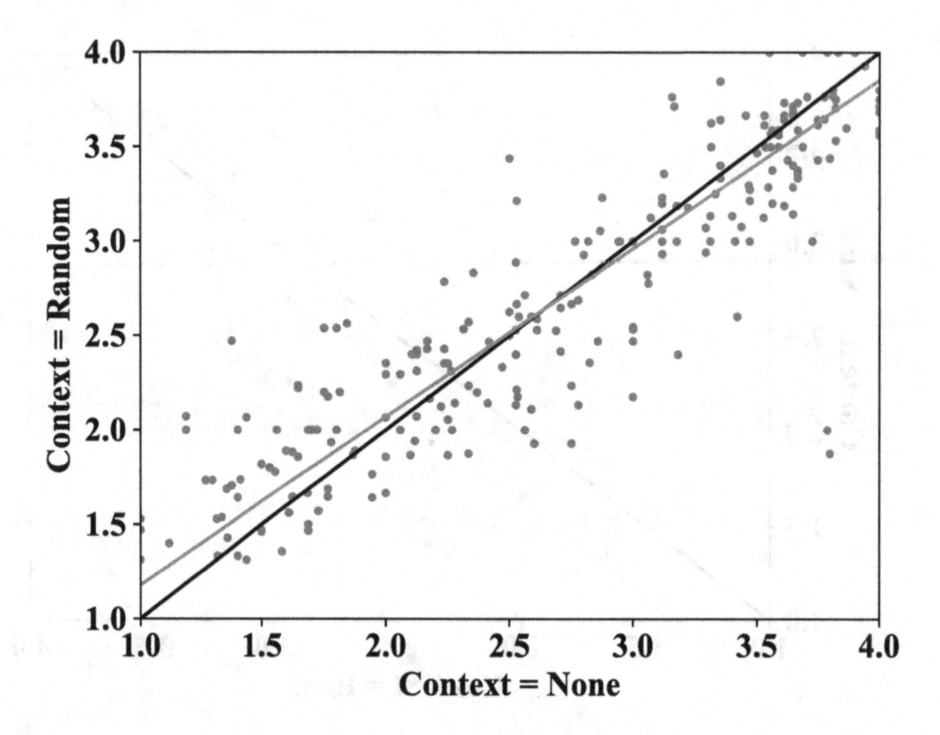

Figure 4.10 Lau's TLS regression for random context vs no context ratings.

4.2 gives the details of the models that Lau et al. (2020) test on their annotated context and non-context sets.

They use Lau et al. (2015) and Lau, Clark, and Lappin (2017)'s three scoring functions, and they add an additional one, *PenLP*, which is a length penalty. Its parametric value $\alpha = 0.8$ is set experimentally, on the basis of work in machine translation. Lau et al. (2020)'s scoring functions are shown in Table 4.3.

Lau et al. (2020) compute two human performance estimates to serve as upper bounds on accuracy of a model. The first, $\mathtt{ub}_1$, is Lau et al. (2015) and Lau, Clark, and Lappin (2017)'s one-vs-rest annotator correlation, discussed in Chapter 3. They select a random annotator's rating, and compare it to the mean rating of the rest, using Pearson's r. They repeat this procedure for a 1000 trials to get a robust estimate of the mean correlation. $\mathtt{ub}_2$ is a half-vs-half annotator correlation, where for

TABLE 4.2 Model Architectures, Parameters, Size, and Training Corpora

Model	Configuration			Training Data			
	Architecture	Encoding	#Param.	Casing	Size	Tokenisation	Corpora
lstm	RNN	Unidir.	60M	Uncased	0.2GB	Word	Wikipedia
tdlm	RNN	Unidir.	80M	Uncased	0.2GB	Word	Wikipedia
gpt2	Transformer	Unidir.	340M	Cased	40GB	BPE	WebText
$\texttt{bert}_{\texttt{cs}}$	Transformer	Bidir.	340M	Cased	13GB	WordPiece	Wikipedia, BookCorpus
$\texttt{bert}_{\texttt{ucs}}$	Transformer	Bidir.	340M	Uncased	13GB	WordPiece	Wikipedia, BookCorpus
xlnet	Transformer	Hybrid	340M	Cased	126GB	Sentence-Piece	Wikipedia, BookCorpus, Giga5 ClueWeb, Common Crawl

TABLE 4.3 Lau et all. (2020)'s Sentence Acceptability Scoring Functions

Scoring Function	Equation
LogProb	$\log P_m(s)$
Mean LP	$\dfrac{\log P_m(s)}{\lvert s\rvert}$
PenLP	$\dfrac{\log P_m(s)}{((5+\lvert s\rvert)/(5+1))^{\alpha}}$
NormLP	$-\dfrac{\log P_m(s)}{\log P_u(s)}$
SLOR	$\dfrac{\log P_m(s) - \log P_u(s)}{\lvert s\rvert}$

$P(s)$ is the sentence probability, computed using either the uni-prob or bi-prob formula, depending on the model, $P_u(s)$ is the sentence probability estimated by a unigram language model, and $\alpha = 0.8$.

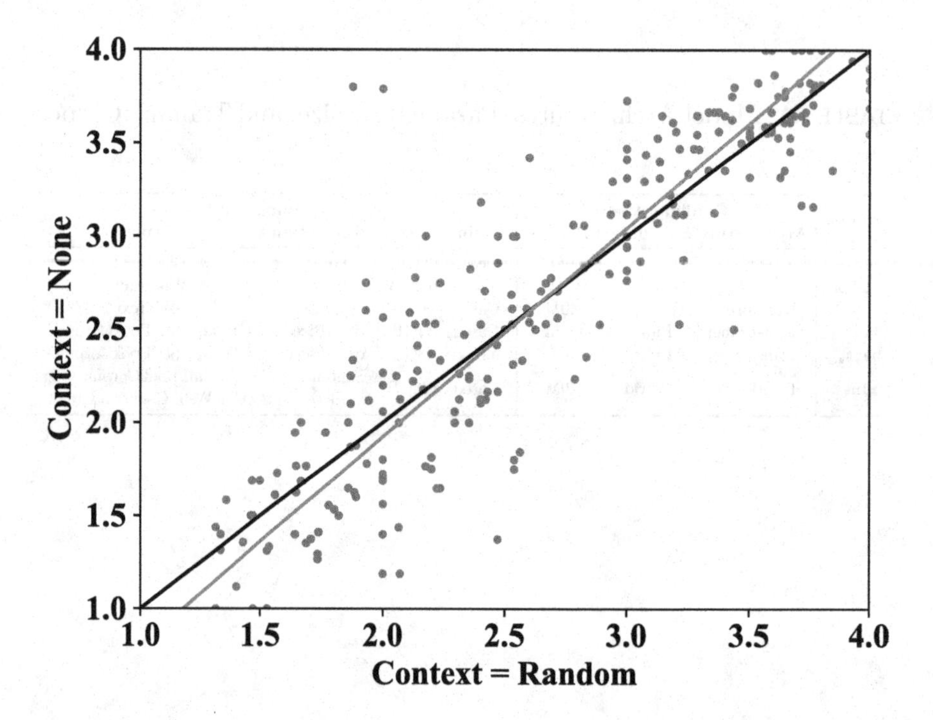

Figure 4.11 Lau's TLS regression for no context vs random context ratings.

each sentence they randomly split the annotators into two groups, and compare the mean ratings between the groups.

Lau et al. (2020) present model performance for the annotation sets in which outlier ratings ($\geq$2 standard deviations) have been removed. This filtering does not significantly affect the model accuracy scores, but it does increase the simulated human upper bound correlations. For completeness, they present the upper bound one-vs-rest correlations for both outlier filtered ($\texttt{ub}_1$,$\texttt{ub}_2$), and outlier unfiltered ($\texttt{ub}_1^{\varnothing}$,$\texttt{ub}_2^{\varnothing}$) test sets.

The results of their model prediction experiments are given in Tables 4.4–4.6, for null ($\texttt{h}^{\varnothing}$), real ($\texttt{h}^{+}$), and random ($\texttt{h}^{-}$) contexts, respectively. M^{+} indicates that the model M was tested with context input, and $M^{\varnothing}$ is for M tested without context input at test time. `bert` is subscripted to show if it was tested on text with, or without uppercase spelling preserved. `xlnet` is indexed for unidirectional, or bidirectional design.

TABLE 4.4 Lau et al. (2020)'s Model Performance for Null Contexts

Rtg	Encod.	Model	*LogProb*	*Mean LP*	*PenLP*	*NormLP*	*SLOR*
$h^{\varnothing}$	Unidir.	lstm$^{\varnothing}$	0.29	0.42	0.42	0.52	**0.53**
		lstm^{+}	0.30	0.49	0.45	0.61	**0.63**
		tdlm$^{\varnothing}$	0.30	0.49	0.45	0.60	**0.61**
		tdlm^{+}	0.30	0.50	0.45	0.59	**0.60**
		gpt2$^{\varnothing}$	0.33	0.34	**0.56**	0.38	0.38
		gpt2^{+}	0.38	0.59	0.58	**0.63**	0.60
		xlnet$^{\varnothing}_{uni}$	0.31	0.42	0.51	0.51	**0.52**
		xlnet$^{+}_{uni}$	0.36	0.56	0.55	0.61	**0.61**
	Bidir.	bert$^{\varnothing}_{cs}$	0.51	0.54	**0.63**	0.55	0.53
		bert$^{+}_{cs}$	0.53	0.63	**0.67**	0.64	0.60
		bert$^{\varnothing}_{ucs}$	0.59	0.63	**0.70**	0.63	0.60
		bert$^{+}_{ucs}$	0.60	0.68	**0.72**	0.67	0.63
		xlnet$^{\varnothing}_{bi}$	0.52	0.51	**0.66**	0.53	0.53
		xlnet$^{+}_{bi}$	0.57	0.65	**0.73**	0.66	0.65
	—	ub$_1$/ ub$^{\varnothing}_1$			0.75 / 0.66		
		ub$_2$/ ub$^{\varnothing}_2$			0.92 / 0.88		

TABLE 4.5 Lau et al. (2020)'s Model Performance for Real Contexts

Rtg	Encod.	Model	*LogProb*	*Mean LP*	*PenLP*	*NormLP*	*SLOR*
h^{+}	Unidir.	lstm$^{\varnothing}$	0.29	0.44	0.43	**0.52**	**0.52**
		lstm^{+}	0.31	0.51	0.46	**0.62**	**0.62**
		tdlm$^{\varnothing}$	0.30	0.50	0.45	**0.59**	**0.59**
		tdlm^{+}	0.30	0.50	0.46	**0.58**	**0.58**
		gpt2$^{\varnothing}$	0.32	0.33	**0.56**	0.36	0.37
		gpt2^{+}	0.38	0.60	0.59	**0.63**	0.60
		xlnet$^{\varnothing}_{uni}$	0.30	0.42	0.50	0.49	**0.51**
		xlnet$^{+}_{uni}$	0.35	0.56	0.55	**0.60**	**0.61**
	Bidir.	bert$^{\varnothing}_{cs}$	0.49	0.53	**0.62**	0.54	0.51
		bert$^{+}_{cs}$	0.52	0.63	**0.66**	0.63	0.58
		bert$^{\varnothing}_{ucs}$	0.58	0.63	**0.70**	0.63	0.60
		bert$^{+}_{ucs}$	0.60	0.68	**0.73**	0.67	0.63
		xlnet$^{\varnothing}_{bi}$	0.51	0.50	**0.65**	0.52	0.53
		xlnet$^{+}_{bi}$	0.57	0.65	**0.74**	0.65	0.65
	—	ub$_1$/ ub$^{\varnothing}_1$			0.73 / 0.66		
		ub$_2$/ ub$^{\varnothing}_2$			0.92 / 0.89		

TABLE 4.6 Lau et al. (2020)'s Model Performance for Random Contexts

Rtg	Encod.	Model	*LogProb*	*Mean LP*	*PenLP*	*NormLP*	*SLOR*
h^-	Unidir.	$\texttt{lstm}^{\varnothing}$	0.28	0.44	0.43	**0.50**	**0.50**
		$\texttt{lstm}^{-}$	0.27	0.41	0.40	**0.47**	**0.47**
		$\texttt{tdlm}^{\varnothing}$	0.29	0.52	0.46	**0.59**	0.58
		$\texttt{tdlm}^{-}$	0.28	0.49	0.44	**0.56**	0.55
		$\texttt{gpt2}^{\varnothing}$	0.32	0.34	**0.55**	0.35	0.35
		$\texttt{gpt2}^{-}$	0.30	0.42	**0.51**	0.44	0.41
		$\texttt{xlnet}_{\texttt{uni}}^{\varnothing}$	0.30	0.44	**0.51**	0.49	0.49
		$\texttt{xlnet}_{\texttt{uni}}^{-}$	0.29	0.40	**0.49**	0.46	0.46
	Bidir.	$\texttt{bert}_{\texttt{cs}}^{\varnothing}$	0.48	0.53	**0.62**	0.53	0.49
		$\texttt{bert}_{\texttt{cs}}^{-}$	0.49	0.52	**0.61**	0.51	0.47
		$\texttt{bert}_{\texttt{ucs}}^{\varnothing}$	0.56	0.61	**0.68**	0.60	0.56
		$\texttt{bert}_{\texttt{ucs}}^{-}$	0.56	0.58	**0.66**	0.57	0.53
		$\texttt{xlnet}_{\texttt{bi}}^{\varnothing}$	0.49	0.48	**0.62**	0.49	0.48
		$\texttt{xlnet}_{\texttt{bi}}^{-}$	0.50	0.51	**0.64**	0.51	0.50
	—	$\texttt{ub}_1$/ $\texttt{ub}_1^{\varnothing}$			0.75 / 0.68		
		$\texttt{ub}_2$/ $\texttt{ub}_2^{\varnothing}$			0.92 / 0.88		

The bidirectional models significantly outperform the unidirectional models across all three context types, when *PenLP*, rather than *SLOR* is the scoring function. This suggests that large lexical embeddings and bidirectional context training render normalisation by word frequency unnecessary. Model architecture rather than size is the decisive factor governing performance. `bert` and `xlnet` approach estimated individual human performance, as specified by $\texttt{ub}_1$, on the prediction of sentence acceptability task for the three context sets. They surpass it for $\texttt{ub}_1^{\varnothing}$ on the null and real context sets.

One might suggest that round-trip MT introduces a systematic bias into the types of infelicities that appear in the Lau et al. (2020) test sets, which could influence the performance of their models. To control for such a possible bias they test the bidirectional transformers, with *PenLP*, on the test set of AMT annotated Adger examples, discussed in Chapter 3. The three bidirectional model scores, with *PenLP*, are: $\texttt{gpt2} = 0.45$, $\texttt{bert}_{\texttt{cs}} = 0.53$, and $\texttt{xlnet}_{\texttt{bi}} = 0.58$. While these scores are lower than those for the round-trip MT test sets, they indicate a strong correlation with human judgements. It is important to note that they are achieved for an out of domain task. The models are trained on naturally occurring text, but they are tested on artificially constructed examples. As we observed in Chapter 3, the linguists' examples are, in general, much shorter than the sentences in the models' training corpora. In fact they are, on average, less than seven words. The difference between the training and test corpora is a significant factor in determining a model's performance on the sentence acceptability task in this case.

4.5 SUMMARY AND CONCLUSIONS

In this chapter, I have looked at recent work on the sentence acceptability task in which sentences are crowd source annotated both out of context, and embedded in different types of document contexts. The first set of experiments compared null to real document contexts, and they tested two types of LSTM LM on the prediction task. One is a simple LSTM, while the other incorporated a topic model. The latter conditioned the prediction of a word in a sequence on both the topic of the sentence and the preceding words. The topic model enhanced LSTM outperformed the simple LSTM, and the addition of document context prefix as test sentence input improved the correlations for both types of LSTM.

Linear regression on the two annotations sets revealed a puzzling compression effect, in which ratings for sentences assessed in context are raised at the bottom end of the scale, but lowered at the higher extreme. This effect was also observed in the unrelated task of rating paraphrase candidates in, and out of context. Predicting acceptability for mean in context ratings is more difficult than the out of context case. This seems to be due to the fact that the judgements are pushed closer to together towards the centre of the rating scale, rendering them less separable.

The second set of experiments that I discussed annotated sentences for null, real, and random contexts, providing three distinct test sets. Detailed analysis of these dataset using both linear and total least square regression shows that the compression effect observed in the earlier work is a real property of the data. Testing the datasets for statistical significance of this effect indicates that both cognitive load and discourse coherence are involved in the in-context ratings. Processing context information induces a cognitive load for humans, which creates a compression effect on the distribution of acceptability ratings. This effect is present in both real and random context sets. If the context is relevant to the sentence, a discourse coherence effect uniformly boosts sentence acceptability. This factor is present only in the real context set.

The second set of experiments tested a variety of DNN LMs, which included the `lstm` and `tdlm` used in the first experiment, and three transformer models, `gpt2`, `bert`, and `xlnet`. The role of case in spelling, and document context input at test time was also considered. The bidirectional transformers outperformed the unidirectional models on the sentence acceptability prediction task. The best bidirectional models approached estimated individual human performance on this task. These models did almost as well for real context ratings as for null context judgements. Random contexts reduced the models' performance more significantly, but even in this case the level of correlation with mean human ratings was robust.

The bidirectional models were tested on AMT annotated Adger sentences to control for the possibility of MT induced bias. While their scores for this set were lower than for the annotated round-trip MT Wikipedia test sets, they remained robust and significant, particularly in view of the fact that this was a strongly out of domain experiment. Bidirectional transformers offer promising models for performing complex NLP tasks that require substantial amounts syntactic and semantic knowledge.

CHAPTER 5

Cognitively Viable Computational Models of Linguistic Knowledge

5.1 HOW USEFUL ARE LINGUISTIC THEORIES FOR NLP APPLICATIONS?

Syntactic and semantic theories offer formal representations of linguistic structure and interpretation, respectively. They aim to express central properties of form and meaning that humans make use of in interpreting the sentences of their languages. If these theories are formally explicit, it is possible to incorporate their principles into computational models of sentence processing. It is reasonable to expect that, to the extent that such theories succeed in capturing core properties of natural language, linguistically informed computational models will show better performance across relevant NLP tasks than systems which do not make use of these theories. In previous chapters, we saw that in general DNNs that incorporate syntactic biases, either in their input data, or their architecture, do not significantly outperform corresponding models in which this bias is absent, across a range of NLP tasks. Let's return to this comparison.

Tai, Socher, and Manning (2015) construct two Tree-LSTM models, one of which encodes dependency trees, and the other constituency trees. Each model produces its hidden state from an input vector and a set of hidden states corresponding to children of the tree node being processed at that point. They apply each model to two tasks: sentiment classification and semantic relatedness. Tai et al. (2015) claim that one of these models outperforms non-tree DNN baselines on (versions) of both tasks. They train their models on the Stanford Sentiment Treebank (Socher et al., 2013), and test them on a subset of this corpus for both five category and binary classification. Their best model, the Constituency Tree-LSTM (with tuned Glove vectors) outperforms the baseline systems on the five category task, with a score of 51.0 accuracy, and it achieves 88.0 on the binary task. The best non-tree LSTM, a Bidirectional LSTM, achieves 49.1 on the five category task, and 87.5 on the binary. Moreover, a (non-tree) multi-channel CNN (Kim, 2014) outperforms both Tree-LSTMs on the binary task, with an accuracy score of 88.1.

For semantic relatedness Tai et al. (2015) train and test their models on the SICK dataset (Marelli et al., 2014), which is annotated by human evaluators. Their Dependency Tree-LSTM scores highest on this task, with a Pearson correlation of 0.8676, and a Spearman correlation of 0.8083. All of the non-Tree LSTMs are above 0.85 on the Pearson metric, and three of the four are above 0.79 on the Spearman (the fourth is at 0.7896). The difference in performance between the Tree- and Non-Tree LSTMs on both tasks is marginal. It is not clear to what extent encoding tree structure in an LSTM improves its handling of either sentiment classification or semantic relatedness.

As we saw in Chapter 2, Bernardy and Lappin (2017) find that training an LSTM on an impoverished lexicon in which POS tags highlight the subject-verb agreement relation, degrades the performance of the model relative to one trained on a richer lexicon. We also observed that, as Williams, Drozdov, and Bowman (2018) show, Choi, Yoo, and Lee (2018)'s latent tree RNN outperforms other systems on two NL inference tasks, but the scores of the non-tree LSTM are not far from those of this model. Latent tree RNNs generate shallow parses, which are not consistent, and do not correlate with theoretically motivated constituency structures.

In Chapter 3, we saw that enriching an LSTM with syntactic or semantic tags, or full dependency trees, degrades its performance on the sentence acceptability task (Ek, Bernardy, & Lappin, 2019).

McCoy, Frank, and Linzen (2020) compare RNNs, LSTMs, and GRUs which incorporate parse tree structure in their respective architectures, with sequential non-tree versions of these DNNs, on two syntactic tasks. For question formation they test which of these models identifies moving the main auxiliary verb, rather than moving the first auxiliary in a sequence, in a generalisation test set.

5.1(a) Don't my yaks that do read giggle?

(b) *Do my yaks that read don't giggle?

For agreement they test the models on the subject vs the most recent NP as controller of the main verb.

5.2(a) My zebra by the yaks swims.

(b) *My zebra by the yaks swim.

McCoy et al. (2020) report that when they train their DNNs on data that contain only ambiguous examples compatible with either rule, the tree-DNNs generalise correctly to unambiguous cases, but the sequential DNNs do not. However, when unambiguous instances of an operation are included in the training data, the sequential and the tree-DNNs perform comparably, with both achieving over 90% accuracy. These results are hardly surprising, given that the tree models incorporate the parse structure into their architecture, and the non-tree models can only learn the correct forms if they are exposed to them in training. It is unclear that these experiments have any consequences for language acquisition, given that there is substantial evidence that human learners are exposed to significant amounts of disambiguating data and reinforcement correction.[1]

In Chapter 2 we discussed Hewitt and Manning (2019)'s supervised squared L2 distance and depth probe for dependency tree structure in DNN sentence vectors. They report that BERT and ELMO, but not their baseline systems, predict unlabelled dependency parse trees, with a high degree of accuracy. This result suggests that these models encode parse structure information in the distributed representations of their lexical embeddings. However, it is not clear that these structures are unambiguously present in the models' output sentence vectors. It is, at least in principle, possible that a different supervised probe would reveal entirely distinct parse structures in these vectors. Additional work is required to rule out this possibility.

Kuncoro et al. (2020) adapt the technique of knowledge distilling that Kuncoro, Dyer, Rimell, Clark, and Blunsom (2019) use for LSTMs,

[1]See A. Clark and Lappin (2011) for discussion and references.

to induce a syntactic bias in BERT. They employ right to left, and left to right RNNG LMs to estimate the trees, and word probabilities, in the right and left contexts of BERT's training data. Kuncoro et al. (2020) then use the combined RNNG LM to supervise the predictions of BERT. They apply both the syntactic knowledge distilled version of BERT (KD BERT) and non-distilled BERT to six NLP tasks requiring syntactic information, and to the GLUE benchmark (Wang et al., 2018), a suite of eight natural language understanding tasks.

In five of the six tasks that Kuncoro et al. test, KD BERT outperforms non-KD BERT by less than 1%. In the sixth task, CCG super tagging, it scores 1.32% higher. Non-KD BERT narrowly outperforms KD BERT on an average of eight tasks for GLUE, with 80.3% to 80%.[2] These results are consistent with the pattern that we observed in our discussion in Chapter 2 of Kuncoro et al. (2019)'s Syntax-Aware LSTM. Although Kuncoro et al. (2020) claim that induced syntactic bias improves the performance of DNN LMs, their results across a wide range of NLP tasks suggest that this bias does not significantly increase accuracy.

5.2 MACHINE LEARNING MODELS VS FORMAL GRAMMAR

In Chapters 3 and 4, we saw that human sentence acceptability judgements are consistently gradient, both at the individual and the aggregate level. LCL use acceptability rather than grammaticality in their crowdsource experiments, because the former is directly observable. By contrast, grammaticality is a theoretical property, and so it is not directly accessible. They seek to avoid biasing the judgements of human annotators with theoretical or prescriptive commitments that they may hold. Theoretical linguists have generally appealed to speakers' acceptability intuitions to motivate their claims.

We saw that deep neural language models achieve encouraging results in predicting mean human acceptability ratings, both in and out of document context. Bidirectional transformers approach estimated human performance for this task on test sets derived by round trip MT on Wikipedia text. They approach, or surpass human performance for this task (when performance is estimated by a one-vs-rest metric unfiltered for outlier judgements). The bidirectional transformer models also

[2]I discuss the Combinatory Categorial Grammar (CCG) model of syntax, and an implemented CCG parser, in Chapter 6.

perform robustly on out of domain prediction of mean human judgements for a linguists' example test set.

Classical formal grammars specify recursive definitions of the set of well-formed sentences in a language. They are binary decision procedures for membership in this class, and so they cannot, in themselves, accommodate gradience. Classical binary theories of grammar must consign gradience to external processing and performance factors. This approach is, in principle, plausible but it must formulate a precise, integrated theory of grammar and processing that predicts the observed phenomenon in detail, to have any explanatory content. To date, no such account has been forthcoming.

Sprouse, Yankama, Indurkhya, Fong, and Berwick (2018) argue that Lau, Clark, and Lappin (2017)'s models capture gradience in human acceptability ratings at the cost of accuracy in binary classification of sentences as acceptable or unacceptable. They train Lau, Clark, and Lappin (2017)'s RNN on the BNC, and they test it, with $SLOR$, on three corpora. These corpora include

i. Sprouse, Schütze, and Almeida (2013)'s set of 150 sentence pairs (good and bad) from *Linguistic Inquiry* articles (LI);

ii. Adger (2003)'s example pairs; and

iii. 120 permutations of the words in *Colorless green ideas sleep furiously* (CGI).

Sprouse et al. (2018) report that Lau, Clark, and Lappin (2017)'s RNN + $SLOR$ achieves Pearson correlations of 0.36 for the mean human ratings of the LI test set, 0.55 for Adger's set, and 0.44 for CGI. They then use the RNN as a binary classifier for the LI and Adger sets, comparing its performance with that of what they describe as a "binary grammar". Their binary grammar is a measure of the Pearson correlation between the linguists' judgements, reported in the LI articles and Adger's textbook, with the mean crowd source acceptability ratings of these sentences. While the Pearson r scores of the RNN are 0.4 for LI and 0.51 for Adger, Sprouse et al. (2018)'s binary grammar metric achieves 0.71 for the former and 0.87 for the latter. They claim that LCL's RNN performs badly in binary acceptability classification in comparison to their binary grammar system.

Lappin and Lau (2018) point out that Sprouse et al. (2018)'s defence of binary grammar as a classifier for sentence acceptability is without force. The "binary grammar metric" which they use as a standard of comparison is neither a grammar nor a model. Instead, it is a version of the one-vs-rest correlation for estimating an upper bound on any model's expected performance. Sprouse et al. (2018) suggest that it is an idealised, if unspecified, categorical grammar from which the linguists' judgements are derived. To assume such a grammar without formulating it is entirely circular, as its existence is the question at issue.

Warstadt, Singh, and Bowman (2019) assembled a Corpus of Linguistic Acceptability (CoLA), a set of 10,657 linguists' sentences labelled for grammaticality/ungrammaticality. They extend it to include out of domain sentences randomly selected from syntax textbooks and research articles. They use five linguistics PhD students to rate a subset of 200 sentences of CoLA for binary acceptability value, and they find that the majority annotator scores diverge from the linguists' annotations for 13% of the subcorpus. This is a comparatively high rate of divergence, which raises the question of the reliability of linguists' grammaticality judgements as evidence for syntactic theories.[3]

Warstadt et al. (2019) do semi-supervised learning for a variety of LSTM and transformer neural LMs by first training them on the sentences of the BNC and ill-formed variants of these sentences derived by permutation. They use rich pre-trained word embeddings in this part of the training process. They then transfer the sentence vectors obtained, to train a binary classifier on their linguists examples for part of CoLA. They test their models, and Lau, Clark, and Lappin (2017)'s RNN (with *SLOR* and the lexical unigram scoring functions), on the remainder of the CoLA corpus.

Warstadt et al. (2019) report that their models generally outperform Lau, Clark, and Lappin (2017)'s RNN. Given the supervised training of these models on CoLA as binary classifiers, and the power of some of their transformers, this is not a surprising result. Interestingly, Lau, Clark, and Lappin (2017)'s RNN is competitive on the out of domain part of the CoLA test set. Also, it outperforms Warstadt et al. (2019)'s models on predicting ratings for three of the five syntactic constructions that they consider.

Hu, Gauthier, Qian, Wilcox, and Levy (2020) test the capacity of a number of neural language models to generalise correctly for a set of syntactic phenomena. The models include, *inter alia*, a vanilla LSTM,

[3]See Gibson and Fedorenko (2013), Sprouse and Almeida (2013), and Gibson, Piantadosi, and Fedorenko (2013) for discussion of this issue.

an RNNG, GPT-2, and GPT-2-XL (both GPT-2 LMs are pre-trained and untuned). The syntactic phenomena which they test are agreement, licensing (negative polarity and reflexive pronouns), garden path effects, the expectation of large syntactic categories, centre embedding, and long distance dependencies (filler-gap structures and cleft verb dependency). The sentences in 5.2 illustrate subject-verb agreement. The following are examples of the other syntactic phenomena.

Licensing: Negative Polarity

5.3(a) The reviewer did **not** raise **any** objections.

(b) *The reviewer raised **any** objections.

Licensing: Reflexive Pronouns

5.4(a) John asked that **we** introduce **ourselves**.

(b) ***We** asked that John introduce **ourselves**.

Garden Path Effects: Main Verb-Reduced Relative Clause

5.5(a) The book **sold** at a discount **was** a first edition.

(b) The book that was **sold** at a discount **was** a first edition.

Syntactic Expectation: Subordinate Clause Requiring a Main Clause

5.6(a) While John read, Mary corrected papers.

(b) *While John read, correct papers.

Centre Embedding:

5.7(a) The student the professor taught graduated.

(b) The student the professor the company hired taught graduated.

Long Distance Dependencies: Filler-Gap Structures

5.8(a) Who did the Chair appoint?

(b) Who did John say that the Committee asked the Chair to appoint?

Pseudo-Clefts:

5.9(a) What John did after coming home is check his email.

(b) What John checked after coming home was his email.

Hu et al. (2020) report that the two GPT-2 transformers outperform the other models on most of the tasks. These models score the highest average accuracy across the tests suites, which is approximately 0.8. They also find that there is no clear correlation between the perplexity of a model, and its performance on the range of tasks that they study. This observation is consistent with Ek et al. (2019)'s results from Chapter 2.

5.3 EXPLAINING LANGUAGE ACQUISITION

Both Sprouse et al. (2018) and Warstadt et al. (2019) suggest that if it is necessary to enrich the training data of ML systems with symbolic features such as part of speech (POS) tags or syntactic trees, then these features will correspond to domain-specific learning biases. They take these biases to be the conditions required for human language acquisition. Sprouse et al. (2018) and Warstadt et al. (2019) identify them with the learning theoretic content of an innate Universal Grammar (UG). In fact this claim is unwarranted.

As we have seen, it is not at all clear that enriching training data with syntactic/semantic markers, or with trees significantly improves the performance of DNN models on most NLP tasks, and, in the case of the sentence acceptability task, such enrichment degrades performance. Even if additional symbolic features were necessary for human level accuracy on NLP tasks, they can be learned from data, without strong domain-specific learning biases. ML systems can efficiently learn POS tags (for example, A. Clark, 2003). Similarly, A. Clark (2015) shows that distributional learning can induce tree structures on strings (strong learning), for a subclass of Context-Free Grammars. A. Clark and Yoshinaka (2014) seek to extend these results to Parallel Context-Free Grammars.

Sprouse et al. (2018) argue from the putative inadequacy of Lau, Clark, and Lappin (2017)'s RNN to the non-viability of ML methods in general, as an approach to modelling language acquisition and linguistic representation. Aside from the fact that their argument concerning the RNN does not go through, it also over reaches in making claims about the full class of ML methods. Lau, Armendariz, Lappin, Purver, and Shu (2020) show that bidirectional transformers approach human performance on the sentence acceptability task.

The history of linguistics and cognitive science is replete with unsound arguments from the limitations of a particular class of models to the non-viability of the entire approach to learning and representation that these models exemplify. I will briefly survey three prominent instances of this kind of argument.

In the first case, Chomsky (1957) observes that simple probabilistic bigram models that use raw frequency counts of lexical items assign nil probability to both 5.10(a) and 5.10(b).

5.10(a) Colourless green ideas sleep furiously.

(b) Furiously sleep ideas green colourlessly.

He concludes that no probabilistic characterisation of grammaticality can succeed, a view that has been widely accepted among theoretical linguists over many years. Pereira (2000) shows that if bigram models are extended to include smoothing for unseen events, and hidden variables for word classes, identified from the data through word distributions, then a bigram model trained on newspaper text assigns a significantly higher probability value to 5.10(a) than to 5.10(b). This is, of course, not to suggest that bigram models can provide an adequate framework for encoding natural language syntax. They clearly do not. However, Pereira's experiment does show that Chomsky's simple argument fails to demonstrate that language models in general are unable to predict relevant differences in likelihood, that correlate with grammatical status, for a large class of sentences.

In the second example, Gold (1967) shows that, given his Identification In the Limit (IIL) learning paradigm, and presentations of positive evidence only, a learner can acquire the class of finite languages, and a finite class of (possibly infinite) languages. However, on IIL with positive evidence only, a learner cannot learn a suprafinite class, which contains the class of finite languages and at least one infinite language. Therefore, none of the language classes of the Chomsky Hierarchy are learnable through induction solely from positive evidence. Some advocates of UG take Gold's results to demonstrate that strong innate, domain-specific constraints on learning are a necessary condition for human language acquisition.[4]

Gold's paradigm relies on a number of highly implausible assumptions concerning the nature of learning, and the evidence available to the language learner. When IIL is replaced by models specified in terms of a more realistic view of the learning process, then it is possible to prove that a much richer class of languages (and of grammars) can be efficiently acquired through data driven induction procedures. These models do not posit strong domain-specific learning biases of the kind encoded in UG. A. Clark and Lappin (2011, 2013) provide detailed discussion of the problems with IIL, and they consider more plausible learning models.

In the third case, Fodor and Pylyshyn (1988) and Fodor (2000) point out the serious limitations in the learning abilities of simple feed forward neural networks. They focus their criticism on the inability of these networks to achieve systematic, rule-like generalisation beyond training data to achieve fully compositional representations of syntactic and semantic properties. They conclude that neural networks in general are incapable of acquiring human level knowledge in most AI applications, particularly in natural language processing. G. Marcus (2001) generalises this criticism to Elman (1990)'s simple RNNs.

Again, the argument involves an unsound inference from the limitations of a particular subclass of ML systems to the non-viability of ML in general as a way of modelling human learning and representation. As we have seen, DNNs, particularly bidirectional transformers, approach, and, in some cases, surpass human performance across a variety of NLP tasks.

However, the issue of systematic generalisation from data to unseen cases remains an important issue in deep learning for NLP and AI. Lake and Baroni (2018) test LSTMs and GRUs for capacity to learn new

[4]See, for example, Crain and Thornton (1998).

actions from training on a corpus of natural language commands. They use a seq2seq encoder-decoder architecture of the sort applied in neural MT. We discussed this architecture in Chapter 2. The encoder takes commands as input, and passes them as context vectors to the decoder, which produces an action sequence. Lake and Baroni (2018) report that their system does well at producing the correct actions for new commands for test cases that vary slightly from those seen in training, but its performance declines substantially as the distance between the test commands and those of the training set increases.

The limitations of Lake and Baroni (2018)'s system would seem to be due, at least in part, to the sort of data that they use for training, and the learning methods that they employ. Hill, Lampinen, et al. (2020) train a neural agent to recognise and implement novel commands in an interactive three-dimensional visual environment that simulates a room with objects of different shapes and colours. They use Reinforcement Learning (RL) to induce rapid zero or one shot learning of new commands.[5] Their agent consists of an LSTM encoder for natural language commands, a CNN for visual input, and a second LSTM that combines language and visual vectors into a multi-modal representation from which it generates an action. Fig 5.1 shows the architecture of their system.

In contrast to Lake and Baroni (2018)'s seq2seq model, Hill, Lampinen, et al. (2020)'s agent does learn to interpret commands that are significantly distinct from the training data, in a robustly systematic way, with limited linguistic evidence. It seems that the use of RL and the richness of the interactive visual environment, are key factors in producing the capacity for systematic generalisation in a DNN. Hill, Tieleman, et al. (2020) demonstrate that a neural agent, trained through RL in an environment of this kind, achieves rapid one shot learning of noun classifier words, which it integrates compositionally into commands.

Work on DNN models for the learning and representation of natural language is still in its infancy. It is reasonable to expect that entirely new types of machine learning architectures will replace current DNNs, and that these may well yield significant gains in modelling ability across a range of linguistic applications. It is necessary for advocates of a categorial grammar, derived from a strong bias UG view of language acquisition, to produce a genuine computational model that provides a

[5]Reinforcement Learning optimises the performance of an ML system through rewarding correct outcomes. See François-Lavet, Henderson, Islam, Bellemare, and Pineau (2018) for an introduction to RL in the deep learning context.

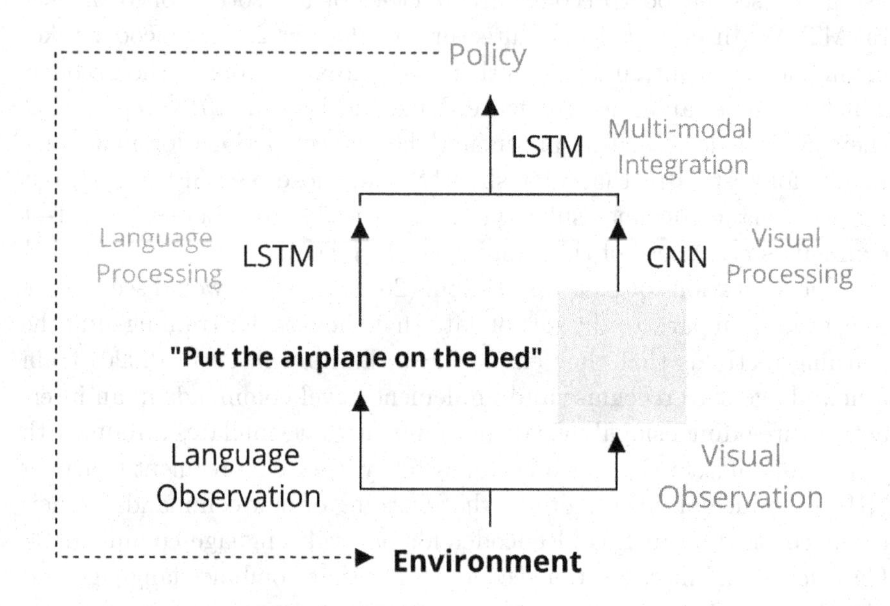

Figure 5.1 Hill et al. 2020's model.

non-trivial classifier for acceptability. Only when such a system is available can we compare a UG approach to ML models for performance in language acquisition, and the handling of NLP applications.

5.4 DEEP LEARNING AND DISTRIBUTIONAL SEMANTICS

Within formal semantics, interpretation consists in mapping the categories of a formal syntax into semantic types that correspond to kinds of denotation in a model theory. Semantic rules compute the denotation of an expression as a relation on the denotations of its syntactic constituents, in tandem with the syntactic operations that generate (or parse) the expression.[6]

Classical semantic theories (Montague, 1974), as well as dynamic (Kamp & Reyle, 1993) and underspecified (Fox & Lappin, 2010) frameworks use categorical type systems. A type T identifies a set of possible denotations for expressions in T. The theory specifies combinatorial operations for deriving the denotation of an expression from the values of its constituents. These theories cannot represent the gradience of semantic properties that is pervasive in speakers' judgements concerning truth, predication, and meaning relations.[7]

Therefore, they suffer from the same difficulty in accommodating the primary linguistic data of gradient speakers' judgements that binary formal grammars exhibit. Even when embedded in probabilistic grammars they are not computationally robust, and they do not yield wide coverage systems. They focus on compositional semantic relations, but do not provide empirically broad or computationally interesting treatments of lexical semantics. It is also not clear how classical semantic theories could be learned from available linguistic data.

Vector Space Models (VSMs) (Turney & Pantel, 2010) offer a fine-grained distributional method for identifying a range of semantic relations among words and phrases. They are constructed from matrices in which words are listed vertically on the left, and the environments in which they appear are given horizontally along the top. These environments specify the dimensions of the model, corresponding to words, phrases, documents, units of discourse, or any other objects for tracking the occurrence of words. As we saw in previous chapters, these co-occurrence matrices are encoded in the word vectors that provide the lexical embeddings of DNNs. They can also include data structures representing extra-linguistic elements, like visual scenes and events. Table 5.1 gives a (contrived) example of a distributional word context matrix that produces vectors with four dimensions.

The integers in the cells of the matrix give the frequency of the word in an environment. A vector for a word is the row of values across the

---

[6]Recently proof theoretic accounts have been proposed for representing meaning in terms of inference rather than denotation (Francez & Dyckhoff, 2010; Richardson, Hu, Moss, & Sabharwal, 2020).

[7]See Sutton (2017) on a probabilistic approach to vagueness in predication. Cooper, Dobnik, Larsson, and Lappin (2015) offer a probabilistic type theory for semantics. Bernardy, Blanck, Chatzikyriakidis, Lappin, and Maskharashvili (2019a, 2019b) propose a Bayesian approach to interpretation and inference.

TABLE 5.1 Word Context Matrix

	context 1	context 2	context 3	context 4
financial	0	6	4	8
market	1	0	15	9
share	5	0	0	4
economic	0	1	26	12
chip	7	8	0	0
distributed	11	15	0	0
sequential	10	31	0	1
algorithm	14	22	2	1

dimension columns of the matrix. The vectors for *chip* and *algorithm* are [7 8 0 0] and [14 22 2 1], respectively.

A pair of vectors from a matrix can be represented geometrically as lines. The smaller the angle between the lines, the greater the similarity of the terms, as measured by their co-occurrence across the dimensions of the matrix. Computing the *cosine* of this angle is a convenient way of measuring the angles between vector pairs. If $\vec{x} = \langle x_1, x_2, ..., x_n \rangle$ and $\vec{y} = \langle y_1, y_2, ..., y_n \rangle$ are two vectors, then $cos(\vec{x}, \vec{y})$ is measured by the following formula.

$$cos(\vec{x}, \vec{y}) = \frac{\Sigma_{i=1}^{n} x_i \cdot y_i}{\sqrt{\Sigma_{i=1}^{n} x_i^2 \cdot \Sigma_{i=1}^{n} y_i^2}}$$

The cosine of $\vec{x}$ and $\vec{y}$ is their inner product formed by summing the products of the corresponding elements of the two vectors, and normalising the result relative to the lengths of the vectors. In computing $cos(\vec{x}, \vec{y})$ it may be desirable to apply a smoothing function to the raw frequency counts in each vector to compensate for sparse data, or to filter out the effects of high frequency terms. A higher value for $cos(\vec{x}, \vec{y})$ correlates with greater semantic relatedness of the terms associated with the $\vec{x}$ and $\vec{y}$ vectors.

VSMs provide highly successful methods for identifying a variety of lexical semantic relations, including synonymy, antinomy, polysemy, and hypernym classes. They also perform very well in unsupervised sense disambiguation tasks. VSMs offer a distributional view of lexical semantic learning. On this approach speakers acquire lexical meaning by estimating the environments (linguistic and non-linguistic) in which the words of their language appear.

VSMs measure semantic distances and relations among words independently of syntactic structure (bag of words). Earlier work sought both to integrate syntactic information into the dimensions of the vector matrices (Padó & Lapata, 2007), and to extend VSM semantic spaces to the compositional meanings of sentences. Mitchell and Lapata (2008) compare additive and multiplicative models for computing the vectors of complex syntactic constituents, and they demonstrate better results with the latter for sentential semantic similarity tasks. These models use simple functions for combining constituent vectors, and they do not represent the dependence of composite vectors on syntactic structure.

Coecke, Sadrzadeh, and Clark (2010) and Grefenstette, Sadrzadeh, Clark, Coecke, and Pulman (2011) propose a procedure for computing vector values for sentences that specifies a correspondence between the vectors and the syntactic structures of their constituents. This procedure relies upon a category theoretic representation of the types of a pregroup grammar (PGG) (Lambek, 2008), which builds up complex syntactic categories through direction-marked function application in a manner similar to a basic categorial grammar. All sentences receive vectors in the same vector space, and so they can be compared for semantic similarity using measures like cosine. A PGG compositional VSM (CVSM) computes the values of a complex syntactic structure through a function that computes the tensor product of the vectors of its constituents, while encoding the correspondence between their grammatical types and their semantic vectors. For two (finite) vector spaces A, B, their tensor product $A \otimes B$ is a vector space constructed from A and B that is constrained only by the relation of bilinearity. A bilinear function from two vector spaces to a third combines the elements of the vectors in each of the two spaces in its domain to generate vectors in its range such that it is linear in each of its argument spaces. For any two vectors $v \in A$, $w \in B$, $v \otimes w$ is the vector consisting of all possible products $v \times w$. Smolensky (1990) uses tensor products of vector spaces to construct representations of complex structures (strings and trees) from the distributed variables and values of the units in a neural network.

PGGs are modelled as *compact closed categories.* A sentence vector is computed by a linear map f on the tensor product for the vectors of its main constituents, where f stores the type categorial structure of the string determined by its PGG representation. The vector for a sentence headed by a transitive verb, for example, is computed according to the equation

$$\overrightarrow{subj\ V_{tr}\ obj} = f(\overrightarrow{subj} \otimes \overrightarrow{V_{tr}} \otimes \overrightarrow{obj}).$$

The vector of a transitive verb V_{tr} could be taken to be an element of the tensor product of the vector spaces for the two noun bases corresponding to its possible subject and object arguments ($\overrightarrow{V_{tr}} \in N \otimes N$). Then the vector for a sentence headed by a transitive verb could be computed as the point-wise product of the verb's vector, and the tensor product of its subject and its object.

$$\overrightarrow{subj\ V_{tr}\ obj} = \overrightarrow{V_{tr}} \odot (\overrightarrow{subj} \otimes \overrightarrow{obj})$$

PGG CVSMs offer a formally grounded and computationally efficient method for obtaining vectors for complex expressions from their syntactic constituents. They permit the same kind of measurement for relations of semantic similarity among sentences that lexical VSMs give for word pairs. They can be trained on a (PGG parsed) corpus, and their performance evaluated against human annotators' semantic judgements for phrases and sentences.

However, they suffer from a crucial conceptual problem, which derives from an empirical difficulty. Although the vector of a complex expression is the value of a linear map on the vectors of its parts, it is not obvious what independent property this vector represents. Sentential vectors do not correspond to the distributional properties of these sentences, as the data is too sparse to estimate distributional vectors for all but a few sentences, across most dimensions. By contrast, DNNs produce sentential vectors by combining input word vectors through a training process controlled by gradient descent and back propagation. They do not, in the general case, represent either sentence frequency or the internal syntactic structure of a sentence. They are generated by optimising the performance of the DNN on a learning task.[8]

[8]Stephen Clark points out that PGG CSVMs could be incorporated into a DNN

VSMs are interesting to the extent that the sentential vectors that they assign are derived from lexical vectors that represent the distributional properties of these expressions. VSMs measure intra-corpus semantic relations. They need to be extended to language-world relations to provide a fully adequate representation of semantic knowledge.

As we saw in Chapter 1, Bahdanau, Cho, and Bengio (2015) construct an encoder-decoder neural translation system, enhanced with attention. The encoder maps the sequence of input word vectors from the source language to a context vector c. The decoder incrementally generates the target sequence by estimating the conditional probability of each target word, given the preceding target words and c. $p(y_i \mid y_1, ..., y_{i-1}, x) = g(y_{i-1}, s_i, c_i)$, where g is a nonlinear (possibly multilayer) function that gives the probability of the target word y_i, s_i is a hidden state of the RNN for time i, and c_i is the context vector for y_i. The RNN is bidirectional in that it searches backwards and forwards through the entire context vector for the input sequence to generate the most likely target words.

Socher, Karpathy, Le, Manning, and Ng (2014), Karpathy and Fei-Fei (2015), Xu et al. (2015) and Vinyals, Toshev, Bengio, and Erhan (2014) apply encoder-decoder architecture to the task of generating descriptions for visual images, and assigning images to descriptions of scenes. Their models are trained on datasets of images annotated with sentences or captions (Flickr 8k, Flickr 30k). The encoder is a CNN that maps sets of pixels in an image into vectors corresponding to visual features. The decoder is an RNN that generates a description from the vector inputs produced by the encoder. The system aligns image features with words and phrases in the way that deep neural MT does for source and target linguistic expressions.

Socher et al. (2014) use dependency parse trees as the input to the decoder RNN of their image description generator. A function g_θ computes the compositional vectors of the parent nodes in the tree. Unlike earlier compositional VSM models the value of this function is learned through back propagation. The dependency tree RNN outperforms constituency tree and bag of words decoder RNNs in generating image descriptions, and in identifying suitable images from descriptions.

---

that was trained to optimise performance on a downstream NLP task. Such a system would be an instance of a DNN with syntactic structure built into its architecture for the purpose of determining the way which input word vectors are combined to generate sentence vectors.

Karpathy and Fei-Fei (2015), Xu et al. (2015), and Vinyals et al. (2014) develop image caption systems that use CNNs to encode images in vectors. They employ RNNs that apply directly to word sequences as decoders. Karpathy and Fei-Fei (2015) use a bidirectional RNN, while Xu et al. (2015) and Vinyals et al. (2014) apply an LSTM RNN. These models can generate descriptions for sub-scenes in an image. They yield better results than Socher et al. (2014)'s system. Fig 5.2 shows the Karpathy-Feifei model for captioning. Fig 5.3 displays an image description from Xu et al. (2015).[9]

In more recent work, transformers have replaced RNNs and CNNs in the encoder-decoder architecture of image captioning systems (He et al., 2020; Herdade, Kappeler, Boakye, & Soares, 2019).

The MT model of image caption generation suggests an approach to semantic interpretation. The encoder component of the model could map an integrated data structure for visual, audio, and textual input representing a situation to vector space features. The decoder would align words and phrases to these features, generating sentences describing the situation, or part of it. Conversely, situations (scenes) could be produced for sentences that correspond to them.

The classical formal semantic program (Davidson, 1967; Montague, 1974) seeks a recursive definition of a truth predicate which entails appropriate truth conditions for each declarative sentence in a language. To the extent that it is successful, a generalised multi-modal MT model would achieve the core part of this program. It would specify suitable correspondences between sentences and sets of situations that the sentences describe. These correspondences are produced not by a recursive definition of a truth predicate, but by an extended DNN language model.

[9] In a talk presented to the Centre for Linguistic Theory and Studies in Probability at the University of Gothenburg on March 11, 2016 (`https://gu-clasp.github.io/events/seminars/2016-03-11/John-Kelleher-Attention/`)
John Kelleher observed that spatial terms like *over* in the Xu et al. (2015) image description are not keyed to visual features of the image. He argues that these terms are generated directly by the RNN language model through the conditional probabilities of the preceding word sequences. Kelleher suggests that it may be possible to ground spatial terms in dynamic image sequences provided by videos, as these offer richer and more salient representations of language-spatial correspondences. See S. Chen, Yao, and Jiang (2019) for a review of recent work on video captioning with DNNs.

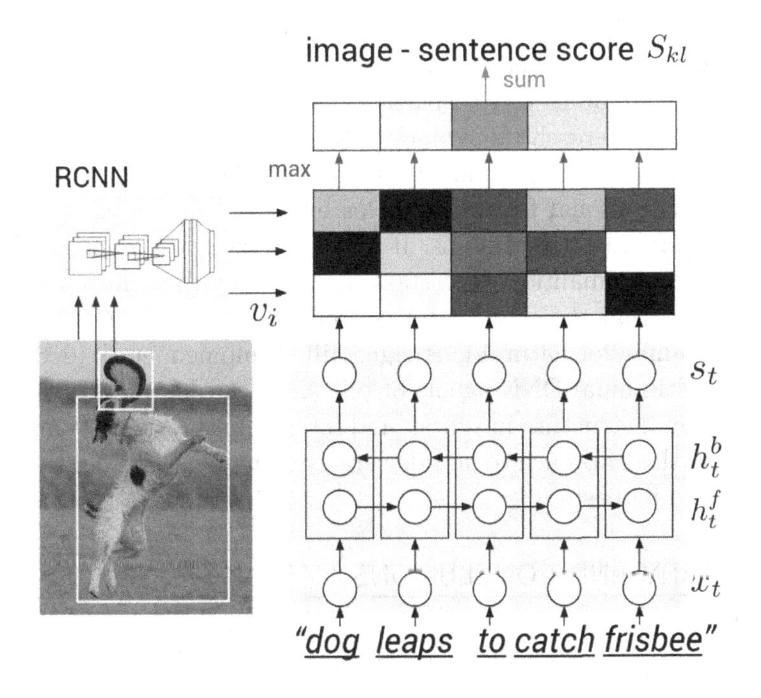

Figure 5.2 Karpathy and Fei-Fei's (2015) model.

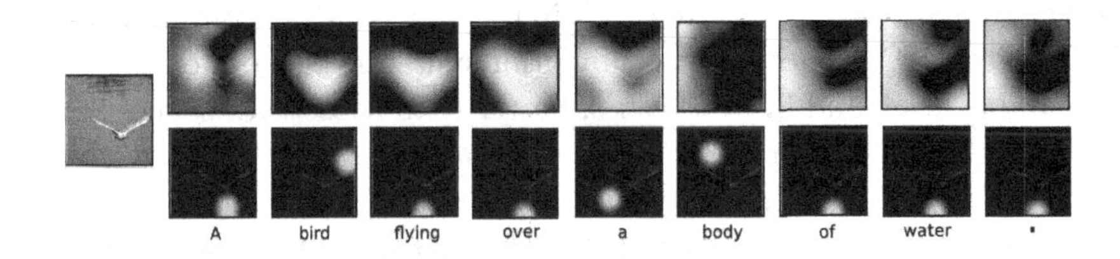

Figure 5.3 Xu et al. (2015) Description of an image.

Classical approaches to semantic interpretation assign semantic values to hierarchical syntactic structures, and they compute the interpretations of sentences by applying combinatorial rules to these values. These approaches provide formally elegant systems, but they have not yielded wide coverage methods for semantic learning and representation. They have also not integrated compositional and lexical meaning in a natural and computationally efficient way.[10]

Multi-modal deep neural MT approaches to image description suggest a model for wide coverage semantic learning and representation that is driven by aligning vector space encodings of linguistic and non-linguistic entities. Hierarchical syntactic structure need not be included in the linguistic input to this model, but it is implicitly represented as part of the distributional feature patterns expressed in the vectors assigned to lexical and phrasal items. If it is successful, the multi-modal deep MT model of semantic interpretation will satisfy the central condition of adequacy that the classical formal semantic program imposes on a theory of meaning for natural language. Hill, Lampinen, et al. (2020)'s interactive multi-modal DNN agent for one shot learning of new words and combinatorial structure in syntax and interpretation suggests a way of generalising this approach to non-declarative sentences, like questions, commands, and requests.

5.5 SUMMARY AND CONCLUSIONS

In this chapter we returned to the comparison of DNNs that incorporate the formalisms of linguistic theories, through annotated training data, architectural design, or mentor induced syntactic bias, with DNNs that do not make use of these formalisms. We asked whether the former significantly outperform the latter on NLP tasks. The evidence on this question remains equivocal. However, in the cases that we considered, we observed that DNNs not informed by prior syntactic bias performed at a comparable level, and in some cases, better than those with such a bias. Hu et al. (2020)'s experiments are particularly noteworthy. They show that two (untuned) GPT-2 transformers are more successful than a variety of sequential models, including an RNNG, in generalising across a variety of syntactic constructions that have been the focus of study in syntactic theory.

One might object to the view that linguistic theories should, in part, be evaluated on their contribution to the performance of machine learning models in solving NLP tasks, on the grounds that these theories are not designed for engineering work. They are intended to explain formal and cognitive properties of natural languages. It is not clear how

[10]Lewis and Steedman (2013) suggest a computational framework for integrating logical compositional representations with distributional semantics.

much substance attaches to such a reply. In other areas of science, engineering tasks depend upon the information provided by theoretical models. Virtually every domain of mechanical, electrical, and aerospace engineering makes crucial use of the principles and equations of theories in physics. Similarly, clinical medicine is heavily dependent upon theoretical research in physiology, neuroscience, genetics, etc. It would be remarkable if natural language were the one domain of science in which accepted theories have no application in engineering.

Linguistic structures of the sort posited by a formal grammar may be implicitly represented in the sentence vectors generated by DNNs. Hewitt and Manning (2019) use a distance and depth probe to identify dependency parse trees in the sentence vector outputs of Elmo and Bert. While this is an intriguing result, it is not clear that such structures are uniquely and unambiguously encoded in transformer sentence vectors. We need additional evidence to rule out the possibility that other supervised probes would identify alternative parse structures corresponding to other grammar formalisms, with equal or greater likelihood. Should such probes emerge, then it would indicate that all of these metrics are finding what they are trained to recognise, in a way that allows for incompatible analyses of the same sentences. Williams, Drozdov, and Bowman (2018)'s work on latent tree RNNs, discussed in Chapter 2, suggest that this sort of pervasive non-uniqueness of parse structures is entirely possible.

We took up Sprouse et al. (2018)'s criticism of Lau, Clark, and Lappin (2017)'s models on the grounds that, although they capture gradience in human sentence acceptability ratings, they underperform in binary classification for grammaticality. We argued that their main argument for this claim lacks any force. It is based on a "categorical grammar metric" which consists of linguists' judgements on the acceptability of the sentences in their test sets. As they do not provide a formal or computational version of the model from which they claim that these judgements are derived, the argument is circular. It presupposes the existence of a binary formal grammar as the representation of human linguistic knowledge, which this is precisely the issue under debate. Traditionally linguists have claimed that linguistic competence is encoded in a formal grammar, while the gradience observed in human sentence acceptability ratings is due to processing and performance effects. In order for this approach to have any explanatory content, it is necessary to formulate precise models of competence, performance, and processing, and to show how their interaction produces the observed data. To the best of my knowledge, no such theory has yet been proposed.

Sprouse et al. (2018) and Warstadt et al. (2019) argue that if additional symbolic features or structural biases are required for machine learning systems to acquire syntactic knowledge effectively, then these represent the elements of a domain-specific innate UG. We saw that this assertion is not well motivated. Work in computational learning theory over the past two decades has demonstrated that rich classes of grammar can be acquired from positive data, by domain general induction procedures.

We observed that Sprouse et al. (2018)'s conclusion that machine learning methods in general cannot produce an adequate model of syntactic learning and representation is an instance of unsound inference from the fact that a member of a class of models is inadequate, to the assertion that the entire class is limited in this way. We looked at three well-known arguments of this type, and we provided evidence that each is a case of over reach. These arguments have been influential in linguistics and cognitive science. It is important to see why they do not hold, and to set them aside.

Finally, we considered distributional semantics and deep learning. We observed that both classical and more recent theories of formal semantics suffer from the same difficulties of formal theories of syntax. They do not yield robust, wide coverage systems of learning and representation, and they are unable to accommodate the gradience that is pervasive in human semantic, as well as syntactic judgements. They also do not integrate lexical semantic information into compositional semantic representations in an interesting or computationally viable way.

We looked at VSMs, which provide the formal basis for lexical embeddings in DNNs, and we briefly considered earlier attempts to compose sentential vectors from distributional lexical vectors. These involved the use of arithmetic functions, or the application of tensor operations in a PGG. While the PGG VSM system is formally elegant, it does not provide a computationally robust framework for learning semantic values of sentences. Moreover, it is not obvious how to interpret the phrasal and sentential vectors that the grammar produces, as they are not empirically grounded in any obvious way.

We concluded with the suggestion that neural image description models might provide a paradigm for developing a viable compositional semantics in which sentence vectors are mapped into multi-modal vector representations of scenes and situations. This approach formulates the problem of semantic interpretation as the task of devising an encoder-decoder sequence-to-sequence MT system that translates sentences into the representations of the situations that they correspond to. Such a system would satisfy the central condition of the formal semantics project, which requires that an adequate semantic theory generate the truth conditions of each well-formed declarative sentence of the language. However, it would fulfil this condition not by means of a recursive definition of the truth predicate of the language, but through a multi-modal neural translation system. Hill, Lampinen, et al. (2020)'s multi-modal interactive RL model offers a natural framework for extending this approach to semantic learning in general, and to the interpretation of non-declarative sentences.

CHAPTER 6

Conclusions and Future Work

6.1 REPRESENTING SYNTACTIC AND SEMANTIC KNOWLEDGE

Throughout the history of linguistics, theorists have developed algebraic systems to represent linguistic knowledge. They have used a variety of grammatical formalisms consisting of rules and conditions to capture the syntactic and semantic properties of natural language. Here is a brief overview of some of these formalisms.

Transformational grammar (Chomsky, 1957, 1965) consists of combinatorial operations to generate phrase structures, and movement to map these structures into permuted forms. In the Principles and Parameters (Chomsky, 1981) and Minimalist (Chomsky, 1995) versions of this approach the grammar of a particular language is derived from a schematic UG through the assignment of values to its parameters.[1]

Head-Driven Phrase Structure Grammar (HPSG) (Pollard & Sag, 1994) analyses phrase structure as attribute-value graphs of typed features. The feature structures of phrases are projected from their lexical heads, and the complements and adjuncts of these heads. Constraints control their combination into phrasal patterns, and the passing of features up through complex graphs by structure sharing operations (an

[1]Johnson and Lappin (1999) provide a detailed critique of earlier formulations of the Minimalist Program. Collins and Stabler (2016) propose a formalisation of the Minimalist approach to syntax. See Lappin and Shieber (2007) and A. Clark and Lappin (2011) for critical discussions of UG in the Principles and Parameters, and the Minimalist frameworks.

abstract form of feature unification). The graph for an expression contains parallel feature paths for its syntactic, semantic, pragmatic, and phonological properties.[2]

Lexical Functional Grammar (LFG) (Bresnan, 2001; Kaplan & Bresnan, 1982) employs phrase structure rules to produce the constituent structure of a sentence, and an untyped feature graph to express its predicate-argument configuration. Meta-rules specify a systematic correspondence between the two representations.

Type logical Categorial Grammars (Moortgat, 1997; Morrill, 1994) are a development of the Lambek calculus (Lambek, 1958). They construct phrases and sentences through the application of functional categories, associated with lexical heads, and logical combinators, to argument categories of the appropriate type. Semantic operations apply in parallel to the categorial construction of an expression. Both their syntactic and their semantic type systems are formulated as typed λ-calculi. The production of the syntactic structure of an expression, and its semantic interpretation, are formulated as co-ordinated logical derivations in each calculus.

Combinatory Categorial Grammar (CCG) (Steedman, 2000) augments the functions of simple categorial grammar with a set of non-logical combinator operations on categories. These increase the expressive power of the grammar, allowing it to raise or change the type of an expression, and to permute the order of constituents in a sentence. Like type logical grammars, CCG runs syntactic and semantic derivations in parallel, with the latter producing typed λ-terms, which can be interpreted model theoretically, as in Montague grammar (Montague, 1974).[3]

Tree Adjoining Grammars (TAGs) (Joshi & Schabes, 1997) generate constituent structure trees through phrase structure rules, and extend them with substitution and adjunction operations. Synchronous TAGs (Shieber & Schabes, 1990) specify systematic correspondences between

[2]Type Theory with Records (TTR) (Cooper & Ginzburg, 2015) uses attribute-value graphs with parallel paths of types, including records, to capture the linguistic properties of utterances. These graphs resemble those of HPSG, but, while HPSG graphs use typed features, TTR employs the full power of the typed λ-calculus, augmented with record types.

[3]In Dynamic Syntax (Kempson, Meyer-Viol, & Gabbay, 2000) syntactic trees are incrementally projected from elements in the sequence of terms in an expression. Each node of the tree is resolved through a type assignment and a λ-term for the expression at that node. This system is designed to capture the dynamic, underspecified, and sequential nature of processing.

TAG trees and the terms of a semantic representation language. Joshi, Shanker, and Weir (1990) show that TAG and CCG are weakly equivalent in generative power, and both are in the class of Mildly Context-Sensitive Grammars (MCSGs).[4]

Although the concept of mild context-sensitivity is not fully defined, the class of MCSGs exhibits two important formal properties. First, they are able to generate cross-serial syntactic dependencies of the kind that Shieber (1985) showed to be instantiated in Swiss German. As these structures are beyond the expressive power of Context-Free languages, it follows that natural languages are at least in the class of Mildly Context-Sensitive languages. Second, while the full class of Context-Sensitive Grammars contains members that generate languages whose upper bound parsing time is exponential, MCSGs only produce languages that can be parsed in polynomial time. Therefore they are a weaker subset of the class of Context-Sensitive Grammars, that extend the expressive power of CFGs in a more restricted way.

Finally, Arc Pair Grammar (APG) (Johnson & Postal, 1981) represents syntactic structures through directed graphs that connect grammatical relations. Conditions on these graphs, and relations between graph pairs, define the set of well-formed sentences in a language. APG adopts a model theoretic, rather than a generative, approach to grammar. Its constraints on graphs are, in effect, restrictions on the set of possible models, whose elements are the expressions of a language. Grammatical sentences are those expressions that exist in all and only the set of possible models defined by the constraints of an APG.[5]

Much valuable linguistic research has been done within these theories of grammar. This work has highlighted central properties of particular languages, and it has drawn attention to interesting cross linguistic patterns of different types. It has also revealed significant formal properties exhibited by classes of natural languages. However, it is not clear that formal grammars offer the most appropriate framework for representing linguistic knowledge. As we have seen in previous chapters, they are unable to express some of the most central features of observed linguistic data. Nor have they supplied the basis for a computationally efficient model of language learning. Exploring alternatives to formal grammars, and other algebraic models, does not entail dismissing the important

[4]Pure categorial grammars containing only functor and arguments types, without additional type constructors or combinatory functions, are CFGs.

[5]See Pullum and Scholz (2001) on the distinction between model theoretic and generative views of grammar.

research that has been conducted within linguistic theory. Instead, it involves taking seriously the possibility that probabilistic machine learning models offer more suitable computational devices for encoding linguistic representations, and for explaining language acquisition.

Some of these grammar formalisms have been used to drive wide coverage parsers. For example, S. Clark and Curran (2007) develop an efficient statistical CCG parser that performs well on parsing metrics. It is trained on a CCG version of the Penn Tree Bank, and it uses a Maximum Entropy supertagger, that assigns CCG lexical category features to the words in a sentence.[6] Although parsing is a natural language processing task, the performance of a parser does not, in itself, provide evidence for the appropriateness of the grammar formalism that it applies, as a system for representing human linguistic knowledge. The parser is evaluated against a gold standard test set annotated by experts who determine how the sentences in this set are to be analysed within the grammatical framework that the parser uses. This metric is not a measure of human performance on a linguistic task, but a record of theoretical judgements by people trained in the application of a system of grammatical analysis.

By contrast, the contribution of a parser to the performance of a machine learning system in solving a cognitively interesting NLP task is relevant to the assessment of the grammar that the parser uses, as a theory of linguistic representation. In previous chapters we have seen that in most of the tasks that we considered, DNNs informed by the syntactic bias of a grammar have not performed significantly better than comparable DNNs which did not incorporate this bias. In many of these tasks, transformers like BERT and GPT-2, applied without fine tuning or knowledge distillation, achieved higher scores than syntactically primed DNNs.

These results do not imply that knowledge of syntax is irrelevant for solving NLP tasks. It is certainly the case that humans recognise and apply syntactic structure in performing these tasks. Rather, the results suggest the possibility that the way in which both DNNs and humans represent syntax is entirely distinct from the formal encoding of combinatorial structures in the algebraic models that have dominated linguistic theory over many decades. We observed in Chapter 2 that the distributed representations of sentence vectors, derived from lexical embeddings, contain syntactic information in the form of large scale cooccurrence patterns.

[6]See Chapter 3, fn 2 for a brief characterisation of Maximum Entropy models.

Hierarchical constituency, dependency, and agreement relations are also implicit in these vectors.

The fact that DNNs can be used as neural language models that generate probability distributions over sequences allows them to accommodate gradience in acceptability judgements in a straightforward way. Formal grammars are not designed to express this property of natural language. This is also a problem for formal semantic models, which do not express degrees of semantic well-formedness. Moreover, unlike grammar driven parsers, most formal semantic theories do not admit of computationally robust, wide coverage implementation.[7] Probability distributions can also be used to capture vagueness in predication. A DNN represents a classifier as applying to objects, relations, or events, to a given degree of likelihood.[8]

We saw in Chapter 2 that it is not possible to reduce acceptability directly to probability, if one invokes a specified probability value as the threshold of acceptability. Adopting such a threshold entails that the number of acceptable sentences in the language is finite, contrary to fact. Using scoring functions to map *LogProb* values into acceptability scores which do not sum to 1 solved this problem. These functions are motivated by the fact that in filtering out the effects of frequency and sentence length they produce better correlation between the distributions of a machine learning model and mean human acceptability ratings.

Formally, it is also possible to avoid the problem by dispensing with a threshold value for acceptability, and treating it as a relational property. On this view, the elements of a set of sentences are more or less acceptable relative to each other. Even if we should use an unfiltered probability distribution to assess relative acceptability, we would not be committed to the existence of a finite set of acceptable sentences, if there is no point that partitions the set into acceptable and unacceptable members.

Many theoretical linguists are acutely uncomfortable with the view that linguistic knowledge is a probabilistic system. Part of this discomfort may well be due to the fact that linguistic theories have traditionally been formulated within symbolic algebraic frameworks, most commonly

[7]Lappin (2015) argues that the set of possible worlds posited in the model theories of many formal semantic accounts, from Montague (1974) through to the present, poses insuperable problems of computational complexity, as this set cannot be efficiently specified or enumerated.

[8]Sutton (2017) and Bernardy, Blanck, Chatzikyriakidis, Lappin, and Maskharashvili (2019a) characterise semantic vagueness in terms of probability distributions.

formal grammars and type theoretic models. This tradition is based on the idea that formal language theory applies to natural languages. To specify a formal language, it is necessary to provide a recursive definition of its expressions. A formal grammar, or a type theory, constitutes a definition of this kind. The observed data of human performance in NLP tasks provides strong prima facie evidence against the view that natural languages are formal objects amenable to such definitions.

In suggesting that it may be more appropriate to represent linguistic knowledge through enriched language models generated by DNNs, we are adopting the view that natural languages are intrinsically indeterminate. Gradience in acceptability judgements, and vagueness in predication are expressions of this underspecification. Using a neural language model to predict acceptability, for example, does not consist in assessing the likelihood that a string is generated by a formal grammar. Hence it is not equivalent to probabilistic parsing. On this approach a neural LM scores sentences for the probability that native speakers of the language will find them more or less acceptable.

Insisting that natural languages are formal languages is reminiscent of Einstein's rejection of quantum mechanics as an ultimate theory of subatomic particles, on the grounds that it undermined a determinate universe governed by universal laws. His famous statement to the effect that God does not play dice with the universe reflects a deep unwillingness to see the world as intrinsically probabilistic in nature.

Regardless of one's views on quantum mechanics, there is an important distinction between this case and the representation of linguistic knowledge as a probabilistic system. Physics is concerned with the laws governing events in the physical world. There is at least an intuitive appeal to Einstein's insistence on a world in which universal laws specify determinate causal relations. By contrast, linguistic knowledge is a human cognitive object. The sort of indeterminacy permitted by probability models seems entirely appropriate for modelling human cognition. The only prior conceptual support available to the formal grammar view of natural language is the tradition of applying formal language theory to natural languages. But, the viability of this enterprise is precisely what is under challenge in this discussion. Most of human knowledge consists in making judgements under conditions of uncertainty, with limited amounts of information. Probability models are designed to capture this process of learning and inference. The experimental work in deep learning in NLP that we have explored in previous Chapters indicates that linguistic knowledge is well suited to representation within this paradigm.

6.2 DOMAIN-SPECIFIC LEARNING BIASES AND LANGUAGE ACQUISITION

The question of what sort of domain-specific bias is required for a DNN to acquire linguistic knowledge is closely connected with the long standing debate between nativists and empiricists on language learning.[9] Where nativists argue for a set of strong domain-specific learning biases, empiricists maintain that natural languages can be learned largely through general inductive inference. It is important to recognise that the empiricist approach posits a rich set of learning mechanisms as part of the architecture of the learner. However, they are not specifically designed to process linguistic data, but drive learning across domains.

Advocates of UG cast the domain-specific bias required for human language acquisition as a schematic grammar. Assigning specific values to the parameters of UG produces the grammar of a particular language, or set of languages. Linguistic nativists regard UG as part of the innate cognitive design of the brain. As we observed in Chapter 5, recent results in computational learning theory demonstrate that, if plausible assumptions are made concerning the data available to the learner, and the nature of the learning process, then expressively powerful classes of grammars can be efficiently inferred from this data, without strong learning biases.

As we saw in previous chapters, there are good reasons for questioning the traditional view that linguistic knowledge is properly represented through algebraic systems like grammars. The neural language models and processing systems that DNNs provide are better suited to capturing the gradience and indeterminacy that are pervasive throughout the subsystems of natural language. We also saw that incorporating syntactic biases into the design, training data, or probability distributions of DNNs did not significantly improve their performance on most of the NLP tasks that we considered. In some cases such biases degraded accuracy.

[9]See Lappin and Shieber (2007) and A. Clark and Lappin (2011) for formulation of the nativist-empiricist controversy in terms of the role of domain specific bias in machine learning. They provide extensive discussion of the issues involved in framing the question in this way.

A criticism commonly raised against DNNs is that they do not generalise robustly beyond their training data, to capture the sorts of combinatorial patterns encoded in grammars, or other symbolic rule-based systems. We noted in Chapter 5 that Lake and Baroni (2018) reported difficulties of this kind in their experiments with training RNNs to recognise new commands. We also saw there that Hill, Lampinen, et al. (2020) and Hill, Tieleman, et al. (2020) show that when an encoder-decoder DNN is trained through RL, with interactive visually grounded data, effective combinatorial generalisation is achieved with limited data, yielding learning of new words and commands in a simulated three-dimensional environment.

These results also show that when DNNs have access to multi-modal interactive data, then learning becomes more efficient in time and quantity of data required. The multi-modal interactive learning paradigm offers a more realistic framework for studying language acquisition. Humans learn their language in a rich non-linguistic environment, where application of expressions to objects and events, and feedback from both peers and mentors on communicative success, are core elements of the acquisition process. It is not surprising that when they are simulated in the training regimen of a DNN, they give rise to a significant improvement in the performance of the system over models trained only on linguistic corpora.

Finally, it is interesting to consider the prospects for constructing a domain general system that handles a variety of tasks, with minimal task specific training. To the extent that it is successful, such a system demonstrates the possibility of using general learning architecture to acquire different sorts of linguistic knowledge. Lu, Goswami, Rohrbach, Parikh, and Lee (2020) construct an encoder-decoder DNN consisting of two BERT components, for linguistic, and for visual input, respectively, with common attentional layers that combine the vectors from each transformer. The system is pre-trained and applied to twelve multi-modal NLP tasks. Lu et al. (2020) report that it outperforms DNNs trained separately for each of these tasks.

The success of domain general DNNs in acquiring the syntactic and semantic knowledge required for a large variety of complex NLP tasks does not resolve the question of whether strong linguistic learning biases are necessary for human language acquisition. However, it does indicate that a computational agent can, in principle, achieve this knowledge efficiently, without incorporating such biases into its internal design.

6.3 DIRECTIONS FOR FUTURE WORK

An important area to be explored in future work is improving the stability and resilience of deep learning NLP systems in the face of noisy data of a kind that humans can filter out. Adversarial testing has been used in image recognition to address the vulnerability of DNNs to small changes in pixel input, which do not affect the classification of a figure. Jia and Liang (2017) show that small additions to suites of test sentences dramatically reduce the accuracy of a text comprehension system. We noted in Chapter 1 that Talman and Chatzikyriakidis (2019) degrade the performance of DNNs trained for natural language inference by substituting alternative lexical items and phrases for some of those in the test set.[10] This technique allows us to identify the points at which a DNN is brittle in its classification of input. We can then work on modifying or revising its learning procedures to accommodate this sort of noise without losing precision.

It will be interesting to explore the possibility of using multi-modal training, with RL, to overcome some of the limitations in DL NLP systems. This line of research would experiment with invoking visual (or other non-linguistic) information to compensate for confusing linguistic input. Reinforcement, and other sorts of interaction could provide additional guidance in filtering this noise in the data.

Expanding training and testing to rich interactive multi-modal input will benefit DL in NLP in general. In Chapter 4, we considered recent experimental work on predicting human acceptability judgements in document contexts. A natural extension of this work would involve testing both human ratings and DNN predictions for sentences presented in non-linguistic contexts, and in dialogue settings.

In Chapter 5 we observed that applying RL in an interactive simulated visual environment permitted DNNs to achieve robust systematic generalisation of combinatorial syntactic and semantic structure. This looks like a promising avenue to explore for improving the performance of NLI systems under adversarial testing. The problems that produced poor compositional semantic generalisation in DNNs trained only on linguistic corpora may be similar, if not identical, to those that cause these systems to under perform on inference classification when trained solely on linguistic data. Humans learn to interpret new sentences, and to infer conclusions from premises, through linguistic interaction in a rich real

[10]See Zhang, Sheng, Alhazmi, and Li (2020) for an extensive overview of adversarial testing methods in deep learning NLP.

world environment. It is reasonable to expect that a cognitively viable computational model of language learning and linguistic representation requires the same sort of input and training.

Finally, we need more intensive collaborative research with neuroscientists and psychologists to ascertain the extent to which the architecture and operation of DNNs that achieve significant success in complex NLP tasks, bear any resemblance to human learning and processing. There are two aspects to this work. The first is discovering whether DL can illuminate the cognitive foundations of human linguistic knowledge, beyond showing how a computational model can achieve some of the linguistic abilities that humans acquire. The second concerns the application of DL to engineering objectives. Language technology will be more effective if it performs its tasks in a way that is familiar to human users. In order to develop this sort of technology it is necessary to have deeper insight into human language learning and processing, so that we can construct DL systems that are informed by this insight.

NLP is now a flourishing area of AI, and the rise of DL methods is, in large part, responsible for the vitality and innovation that we are seeing in this work. DL has greatly expanded the horizons for cognitively interesting computational research on natural language, as well as rapid improvement in language technology across a wide range of applications. It has become a lively point of interface between computer scientists, linguists, and cognitive scientists, sharing methods and results across their disciplines. As the field continues to develop, one hopes that it will sustain a focus on foundational scientific questions. No less important is that it be guided by a firm commitment to ensuring that its technology is used for social benefit.

References

Abzianidze, L., Bjerva, J., Evang, K., Haagsma, H., van Noord, R., Ludmann, P., ... Bos, J. (2017). The parallel meaning bank: Towards a multilingual corpus of translations annotated with compositional meaning representations. *ArXiv preprint arXiv:1702.03964*.

Adger, D. (2003). *Core syntax: A minimalist approach.* United Kingdom: Oxford University Press.

Bahdanau, D., Cho, K., & Bengio, Y. (2015). Neural machine translation by jointly learning to align and translate. *ArXiv, abs/1409.0473*.

Baroni, M., Bernardini, S., Ferraresi, A., & Zanchetta, E. (2009). The wacky wide web: A collection of very large linguistically processed web-crawled corpora. *Language Resources and Evaluation, 43*(3), 209–226.

Berger, J. O. (1985). *Statistical decision theory and Bayesian analysis (2nd ed.).* New York: Springer-Verlag.

Bernardy, J.-P., Blanck, R., Chatzikyriakidis, S., Lappin, S., & Maskharashvili, A. (2019a). Bayesian inference semantics: A modelling system and a test suite. In *Proceedings of the eighth joint conference on lexical and computational semantics (*SEM 2019)* (pp. 263–272). Minneapolis, Minnesota: Association for Computational Linguistics.

Bernardy, J.-P., Blanck, R., Chatzikyriakidis, S., Lappin, S., & Maskharashvili, A. (2019b). Predicates as boxes in Bayesian semantics for natural language. In *Proceedings of the 22nd Nordic conference on computational linguistics* (pp. 333–337). Turku, Finland: Linköping University Electronic Press.

Bernardy, J.-P., & Lappin, S. (2017). Using deep neural networks to learn syntactic agreement. *Linguistic Issues in Language Technology, 15*, 1–15.

Bernardy, J.-P., Lappin, S., & Lau, J. H. (2018). The influence of context on sentence acceptability judgements. In *Proceedings of the 56th annual meeting of the association for computational linguistics (ACL 2018)* (pp. 456–461). Melbourne, Australia.

Bizzoni, Y., & Lappin, S. (2017). Deep learning of binary and gradient judgements for semantic paraphrase. In *IWCS 2017 – 12th international conference on computational semantics – Short papers*.

Bizzoni, Y., & Lappin, S. (2019). The effect of context on metaphor paraphrase aptness judgments. In *Proceedings of the 13th international conference on computational semantics – Long papers* (pp. 165–175). Gothenburg, Sweden.

Blank, I., Balewski, Z., Mahowald, K., & Fedorenko, E. (2016). Syntactic processing is distributed across the language system. *NeuroImage, 127*, 307–323.

Bowman, S. R., Angeli, G., Potts, C., & Manning, C. D. (2015, September). A large annotated corpus for learning natural language inference. In *Proceedings of the 2015 conference on empirical methods in natural language processing* (pp. 632–642). Lisbon, Portugal: Association for Computational Linguistics.

Bowman, S. R., Gauthier, J., Rastogi, A., Gupta, R., Manning, C. D., & Potts, C. (2016, August). A fast unified model for parsing and sentence understanding. In *Proceedings of the 54th annual meeting of the association for computational linguistics (volume 1: Long papers)* (pp. 1466–1477). Berlin, Germany: Association for Computational Linguistics.

Bresnan, J. (2001). *Lexical-functional syntax.* Malden, MA: Blackwell.

Bringsjord, S., & Govindarajulu, N. S. (2020). Artificial intelligence. In E. N. Zalta (Ed.), *The Stanford encyclopedia of philosophy* (Summer 2020 ed.). Metaphysics Research Lab, Stanford University.

Brown, T., Mann, B., Ryder, N., Subbiah, M., Kaplan, J., Dhariwal, P., ... Amodei, D. (2020). Language models are few-shot learners. *ArXiv, abs/2005.14165*.

Causse, M., Peysakhovich, V., & Fabre, E. F. (2016). High working memory load impairs language processing during a simulated piloting task: An ERP and pupillometry study. *Frontiers in Human Neuroscience, 10*, 240.

Chen, D., & Manning, C. (2014). A fast and accurate dependency parser using neural networks. In *Proceedings of the 2014 conference on empirical methods in natural language processing (EMNLP)* (pp. 740–750).

Chen, S., Yao, T., & Jiang, Y.-G. (2019). Deep learning for video captioning: A review. In *Proceedings of the twenty-eighth international joint conference on artificial intelligence, IJCAI-19* (pp. 6283–6290). International Joint Conferences on Artificial Intelligence Organization.

Cho, K., van Merriënboer, B., Gulcehre, C., Bahdanau, D., Bougares, F., Schwenk, H., & Bengio, Y. (2014). Learning phrase representations using RNN encoder–decoder for statistical machine translation. In *Proceedings of the 2014 conference on empirical methods in natural language processing (EMNLP)* (pp. 1724–1734). Doha, Qatar: Association for Computational Linguistics.

Choi, J., Yoo, K. M., & Lee, S.-g. (2018). Learning to compose task-specific tree structures. In *AAAI conference on artificial intelligence.*

Chomsky, N. (1957). *Syntactic structures.* The Hague: Mouton.

Chomsky, N. (1965). *Aspects of the theory of syntax.* Cambridge, MA: The MIT Press.

Chomsky, N. (1981). *Lectures on government and binding.* Dordrecht, Holland: Foris Publications.

Chomsky, N. (1995). *The minimalist program.* Cambridge, MA: The MIT Press.

Clark, A. (2003). Combining distributional and morphological information for part of speech induction. In *Proceedings of the tenth conference on European chapter of the association for computational linguistics* (Vol. 1, pp. 59–66). USA: Association for Computational Linguistics.

Clark, A. (2015). Canonical context-free grammars and strong learning: Two approaches. In *Proceedings of the 14th meeting on the mathematics of language (MoL 2015)* (pp. 99–111).

Clark, A., Fox, C., & Lappin, S. (Eds.). (2010). *The handbook of computational linguists and natural language processing.* Malden, MA and Oxford: Wiley-Blackwell.

Clark, A., & Lappin, S. (2010). Unsupervised learning and grammar induction. In A. Clark, C. Fox, & S. Lappin (Eds.), *The handbook of computational linguists and natural language processing* (pp. 197–220). Malden, MA and Oxford: Wiley-Blackwell.

Clark, A., & Lappin, S. (2011). *Linguistic nativism and the poverty of the stimulus.* Malden, MA and Oxford: Wiley-Blackwell.

Clark, A., & Lappin, S. (2013). Complexity in language acquisition. *Topics in Cognitive Science, 5*(1), 89–110.

Clark, A., & Yoshinaka, R. (2014). Distributional learning of parallel multiple context-free grammars. *Machine Learning, 96*(1–2), 5–31.

Clark, S., & Curran, J. R. (2007). Wide-coverage efficient statistical parsing with CCG and log-linear models. *Computational Linguistics, 33*(4), 493–552.

Coecke, B., Sadrzadeh, M., & Clark, S. (2010). Mathematical foundations for a compositional distributional model of meaning. *Lambek Festschrift, Linguistic Analysis, 36.*

Collins, C., & Stabler, E. (2016). A formalization of minimalist syntax. *Syntax, 19*.

Cooper, R., Dobnik, S., Larsson, S., & Lappin, S. (2015). Probabilistic type theory and natural language semantics. *Linguistic Issues in Language Technology, 10*.

Cooper, R., & Ginzburg, J. (2015). Type theory with records for natural language semantics. In S. Lappin & C. Fox (Eds.), *The handbook of contemporary semantic theory* (pp. 375–407). John Wiley & Sons, Ltd.

Crain, S., & Thornton, R. (1998). *Investigations in universal grammar: A guide to experiments on the acquisition of syntax and semantics.* MIT Press.

Davidson, D. (1967). Truth and meaning. *Synthese, 17*(1), 304–323.

de Marneffe, M.-C., MacCartney, B., & Manning, C. D. (2006, May). Generating typed dependency parses from phrase structure parses. In *Proceedings of the fifth international conference on language resources and evaluation (LREC 06).* Genoa, Italy: European Language Resources Association (ELRA).

Devlin, J., Chang, M.-W., Lee, K., & Toutanova, K. (2019, June). BERT: Pre-training of deep bidirectional transformers for language understanding. In *Proceedings of the 2019 conference of the north American chapter of the association for computational linguistics: Human language technologies, volume 1 (long and short papers)* (pp. 4171–4186). Minneapolis, Minnesota: Association for Computational Linguistics.

Dubhashi, D., & Lappin, S. (2017). AI dangers: Imagined and real. *Communications of the ACM, 60*(2), 43–45.

Dyer, C., Kuncoro, A., Ballesteros, M., & Smith, N. A. (2016, June). Recurrent neural network grammars. In *Proceedings of the 2016 conference of the north American chapter of the association for computational linguistics: Human language technologies* (pp. 199–209). San Diego, CA: Association for Computational Linguistics.

Ek, A., Bernardy, J.-P., & Lappin, S. (2019). Language modeling with syntactic and semantic representation for sentence acceptability predictions. In *Proceedings of the 22nd Nordic conference on computational linguistics* (pp. 76–85). Turku, Finland.

Elman, J. L. (1990). Finding structure in time. *Cognitive Science, 14*(2), 179–211.

Fodor, J. A. (2000). *The mind doesn't work that way: The scope and limits of computational psychology.* MIT Press.

Fodor, J. A., & Pylyshyn, Z. W. (1988). Connectionism and cognitive architecture: A critical analysis. In S. Pinker & J. Mehler (Eds.), *Connections and symbols* (pp. 3–71). Cambridge, MA, USA: MIT Press.

Fox, C., & Lappin, S. (2010). Expressiveness and complexity in underspecified semantics. *Lambek Festschrift, Linguistic Analysis, 36*, 385–417.

François-Lavet, V., Henderson, P., Islam, R., Bellemare, M. G., & Pineau, J. (2018). An introduction to deep reinforcement learning. *Foundations and Trends in Machine Learning, 11*(3–4), 219–354.

Francez, N., & Dyckhoff, R. (2010). Proof-theoretic semantics for a natural language fragment. *Linguistics and Philosophy, 33*(6), 447–477.

Gibson, E., & Fedorenko, E. (2013). The need for quantitative methods in syntax and semantics research. *Language and Cognitive Processes, 28*(1-2), 88-124.

Gibson, E., Piantadosi, S. T., & Fedorenko, E. (2013). Quantitative methods in syntax/semantics research: A response to Sprouse and Almeida (2013). *Language and Cognitive Processes, 28*(3), 229–240.

Gold, E. M. (1967). Language identification in the limit. *Information and Control, 10*(5), 447–474.

Goldberg, Y. (2019). Assessing BERT's syntactic abilities. *ArXiv, abs/1901.05287*.

Gómez-Rodríguez, C., & Vilares, D. (2018). Constituent parsing as sequence labeling. *ArXiv*. (arXiv: 1810.08994)

Goodfellow, I., Bengio, Y., & Courville, A. (2016). *Deep learning*. The MIT Press.

Goodman, J. (2001, 10). A bit of progress in language modeling. *Computer Speech & Language, 15*, 403–434.

Grefenstette, E., Sadrzadeh, M., Clark, S., Coecke, B., & Pulman, S. (2011). Concrete sentence spaces for compositional distributional models of meaning. In *Proceedings of the ninth international conference on computational semantics (IWCS 2011)*.

Gulordava, K., Bojanowski, P., Grave, E., Linzen, T., & Baroni, M. (2018). Colorless green recurrent networks dream hierarchically. In *Proceedings of the 2018 conference of the north American chapter of the association for computational linguistics: Human language technologies, volume 1 (long papers)* (pp. 1195–1205). New Orleans, LA: Association for Computational Linguistics.

He, S., Liao, W., Tavakoli, H., Yang, M., Rosenhahn, B., & Pugeault, N. (2020). Image captioning through image transformer. *ArXiv*, 1–17.

Herdade, S., Kappeler, A., Boakye, K., & Soares, J. (2019). Image captioning: Transforming objects into words. In *Neural information processing systems.*

Hewitt, J., & Manning, C. D. (2019, June). A structural probe for finding syntax in word representations. In *Proceedings of the 2019 conference of the north American chapter of the association for computational linguistics: Human language technologies, volume 1 (long and short papers)* (pp. 4129–4138). Minneapolis, MN: Association for Computational Linguistics.

Hill, F., Lampinen, A. K., Schneider, R., Clark, S., Botvinick, M., McClelland, J. L., & Santoro, A. (2020). drivers of systematicity and generalization in a situated agent. In *8th international conference on learning representations, ICLR 2020, Addis Ababa, Ethiopia, April 26–30, 2020.*

Hill, F., Reichart, R., & Korhonen, A. (2015). SimLex-999: Evaluating semantic models with (genuine) similarity estimation. *Computational Linguistics, 41*, 665–695.

Hill, F., Tieleman, O., von Glehn, T., Wong, N., Merzic, H., & Clark, S. (2020). Grounded language learning fast and slow. *ArXiv.*

Hochreiter, S., & Schmidhuber, J. (1997, 12). Long short-term memory. *Neural Computation, 9*, 1735–1780. doi: 10.1162/neco.1997.9.8.1735

Hu, J., Gauthier, J., Qian, P., Wilcox, E., & Levy, R. (2020, July). A systematic assessment of syntactic generalization in neural language models. In *Proceedings of the 58th annual meeting of the association for computational linguistics* (pp. 1725–1744). Online: Association for Computational Linguistics.

Ito, A., Corley, M., & Pickering, M. J. (2018). A cognitive load delays predictive eye movements similarly during l1 and l2 comprehension. *Bilingualism: Language and Cognition, 21*(2), 251–264.

Jia, R., & Liang, P. (2017). Adversarial examples for evaluating reading comprehension systems. In *Proceedings of the 2017 conference on empirical methods in natural language processing* (pp. 2021–2031). Copenhagen, Denmark: Association for Computational Linguistics.

Johnson, D., & Lappin, S. (1999). *Local constraints vs. economy.* Stanford, CA: CSLI Publications.

Johnson, D., & Postal, P. (1981). *Arc pair grammar.* Princeton, NJ: Princeton University Press.

Joshi, A. K., & Schabes, Y. (1997). Tree-adjoining grammars. In G. Rozenberg & A. Salomaa (Eds.), *Handbook of formal languages* (Vol. 3, pp. 69–124). Berlin, New York: Springer.

Joshi, A. K., Shanker, K. V., & Weir, D. (1990). *The convergence of mildly context-sensitive grammar formalisms* (Tech. Rep.). Philadelphia, PA: Department of Computer and Information Science, University of Pennsylvania.

Józefowicz, R., Vinyals, O., Schuster, M., Shazeer, N., & Wu, Y. (2016). Exploring the limits of language modeling. *ArXiv, abs/1602.02410.*

Kamp, H., & Reyle, U. (1993). *From discourse to logic: Introduction to model-theoretic semantics of natural language, formal logic and discourse representation theory* (Vol. 42). Dordrecht: Springer.

Kaplan, R. M., & Bresnan, J. (1982). Lexical-functional grammar: A formal system for grammatical representation. In J. Bresnan (Ed.), *The mental representation of grammatical relations* (pp. 173–281). Cambridge, MA: MIT Press.

Karpathy, A., & Fei-Fei, L. (2015). Deep visual-semantic alignments for generating image descriptions. *ArXiv.*

Kempson, R., Meyer-Viol, W., & Gabbay, D. (2000). *Dynamic syntax: The flow of language understanding.* Wiley-Blackwell.

Kim, Y. (2014). Convolutional neural networks for sentence classification. In *Proceedings of the 2014 conference on empirical methods in natural language processing (EMNLP)* (pp. 1746–1751). Doha, Qatar: Association for Computational Linguistics.

Klein, D., & Manning, C. D. (2003a). Accurate unlexicalized parsing. In *Proceedings of the 41st annual meeting of the association for computational linguistics* (pp. 423–430). Sapporo, Japan: Association for Computational Linguistics.

Klein, D., & Manning, C. D. (2003b). Fast exact inference with a factored model for natural language parsing. In S. Becker, S. Thrun, & K. Obermayer (Eds.), *Advances in neural information processing systems 15* (pp. 3–10). MIT Press.

Koehn, P., Hoang, H., Birch, A., Callison-Burch, C., Federico, M., Bertoldi, N., ... Herbst, E. (2007). Moses: Open source toolkit for statistical machine translation. In *Proceedings of the 45th annual meeting of the association for computational linguistics companion volume proceedings of the demo and poster sessions* (pp. 177–180). Prague, Czech Republic.

Kuncoro, A., Dyer, C., Hale, J., Yogatama, D., Clark, S., & Blunsom, P. (2018). LSTMs can learn syntax-sensitive dependencies well, but modeling structure makes them better. In *Proceedings of the 56th annual meeting of the association for computational linguistics (volume 1: Long papers)* (pp. 1426–1436). Melbourne, Australia: Association for Computational Linguistics.

Kuncoro, A., Dyer, C., Rimell, L., Clark, S., & Blunsom, P. (2019, July). Scalable syntax-aware language models using knowledge distillation. In *Proceedings of the 57th annual meeting of the association for computational linguistics* (pp. 3472–3484). Florence, Italy: Association for Computational Linguistics.

Kuncoro, A., Kong, L., Fried, D., Yogatama, D., Rimell, L., Dyer, C., & Blunsom, P. (2020). Syntactic structure distillation pretraining for bidirectional encoders. *ArXiv*.

Lake, B., & Baroni, M. (2018). Generalization without systematicity: On the compositional skills of sequence-to-sequence recurrent networks. In J. Dy & A. Krause (Eds.), *Proceedings of machine learning research* (Vol. 80, pp. 2873–2882). Stockholm Sweden: PMLR.

Lambek, J. (1958). The mathematics of sentence structure. *American Mathematical Monthly*.

Lambek, J. (2008). Pregroup grammars and Chomsky's earliest examples. *Journal of Logic, Language and Information*, *17*, 141–160.

Lappin, S. (2015). Curry typing, polymorphism, and fine-grained intensionality. In S. Lappin & C. Fox (Eds.), *The handbook of contemporary semantic theory* (pp. 408–428). John Wiley & Sons, Ltd.

Lappin, S., & Lau, J. H. (2018). Gradient probabilistic models vs categorical grammars: A reply to Sprouse et al. (2018). In *The science of language (Ted Gibson's Blog)*.

Lappin, S., & Shieber, S. (2007). Machine learning theory and practice as a source of insight into universal grammar. *Journal of Linguistics*, *43*, 393–427.

Lau, J. H., Armendariz, C., Lappin, S., Purver, M., & Shu, C. (2020). How furiously can colorless green ideas sleep? Sentence acceptability in context. *Transactions of the Association for Computational Linguistics*, *8*, 296–310. doi: 10.1162/tacl_a_00315

Lau, J. H., Baldwin, T., & Cohn, T. (2017). Topically driven neural language model. In *Proceedings of the 55th annual meeting of the association for computational linguistics (volume 1: Long papers)* (pp. 355–365). Vancouver, Canada.

Lau, J. H., Clark, A., & Lappin, S. (2014). Measuring gradience in speakers' grammaticality judgements. In *Proceedings of the 36th annual conference of the cognitive science society* (pp. 821–826). Quebec City, Canada.

Lau, J. H., Clark, A., & Lappin, S. (2015). Unsupervised prediction of acceptability judgements. In *Proceedings of the joint conference of the 53rd annual meeting of the association for computational linguistics and the 7th international joint conference on natural language processing of the Asian Federation of Natural Language Processing (ACL-IJCNLP 2015)* (pp. 1618–1628). Beijing, China.

Lau, J. H., Clark, A., & Lappin, S. (2017). Grammaticality, acceptability, and probability: A probabilistic view of linguistic knowledge. *Cognitive Science*, *41*, 1202–1241.

Lecun, Y., Kavukcuoglu, K., & Farabet, C. (2010, 05). Convolutional networks and applications in vision. In *ISCAS 2010 - 2010 IEEE international symposium on circuits and systems: Nano-bio circuit fabrics and systems* (pp. 253–256).

Lewis, M., & Steedman, M. (2013). Combined distributional and logical semantics. *Transactions of the Association for Computational Linguistics*, *1*, 179–192.

Linzen, T., Chrupała, G., & Alishahi, A. (Eds.). (2018). *Proceedings of the 2018 EMNLP workshop BlackboxNLP: Analyzing and interpreting neural networks for NLP.* Brussels, Belgium: Association for Computational Linguistics.

Linzen, T., Chrupała, G., Belinkov, Y., & Hupkes, D. (Eds.). (2019, August). *Proceedings of the 2019 ACL workshop BlackboxNLP: Analyzing and interpreting neural networks for NLP.* Florence, Italy: Association for Computational Linguistics.

Linzen, T., Dupoux, E., & Goldberg, Y. (2016). Assessing the ability of LSTMs to learn syntax-sensitive dependencies. *Transactions of the Association for Computational Linguistics*, *4*, 521–535.

Lu, J., Goswami, V., Rohrbach, M., Parikh, D., & Lee, S. (2020). 12-in-1: Multi-task vision and language representation learning. *In 2020 IEEE/CVF conference on computer vision and pattern recognition (CVPR)*, 10434–10443.

Maillard, J., Clark, S., & Yogatama, D. (2019). Jointly learning sentence embeddings and syntax with unsupervised Tree-LSTMs. *Natural Language Engineering*, *25*(4), 433–449.

Marcus, G. (2001). *The algebraic mind: Integrating connectionism and cognitive science.* Cambridge, MA: MIT Press.

Marcus, M. P., Santorini, B., & Marcinkiewicz, M. A. (1993). Building a large annotated corpus of English: The Penn Treebank. *Computational Linguistics*, *19*(2), 313–330.

Marelli, M., Bentivogli, L., Baroni, M., Bernardi, R., Menini, S., & Zamparelli, R. (2014). SemEval-2014 task 1: Evaluation of compositional distributional semantic models on full sentences through semantic relatedness and textual entailment. In *Proceedings of the 8th international workshop on semantic evaluation (SemEval 2014)* (pp. 1–8). Dublin, Ireland: Association for Computational Linguistics.

Marvin, R., & Linzen, T. (2018). Targeted syntactic evaluation of language models. In *Proceedings of the 2018 conference on empirical methods in natural language processing* (pp. 1192–1202). Brussels, Belgium: Association for Computational Linguistics.

McCoy, R. T., Frank, R., & Linzen, T. (2020). Does syntax need to grow on trees? Sources of hierarchical inductive bias in sequence-to-sequence networks. *Transactions of the Association for Computational Linguistics*, *8*, 125–140.

Mikolov, T. (2012). *Statistical language models based on neural networks* (Unpublished doctoral dissertation). Brno University of Technology.

Mikolov, T., Kombrink, S., Deoras, A., Burget, L., & Èernocký, J. (2011). RNNLM – Recurrent neural network language modeling toolkit. In *IEEE automatic speech recognition and understanding workshop.* Hawaii, US.

Mikolov, T., Sutskever, I., Chen, K., Corrado, G., & Dean, J. (2013). Distributed representations of words and phrases and their compositionality. In *Proceedings of the 26th international conference on neural information processing systems – volume 2* (pp. 3111–3119). Red Hook, NY, USA: Curran Associates Inc.

Mitchell, J., & Lapata, M. (2008). Vector-based models of semantic composition. In *Proceedings of ACL-08: HLT* (pp. 236–244). Columbus, Ohio: Association for Computational Linguistics.

Montague, R. (1974). *Formal philosophy: Selected papers of Richard Montague.* New Haven, CT/London, UK: Yale University Press. (Edited with an introduction by R. H. Thomason)

Moortgat, M. (1997). Categorial type logics. In J. van Benthem & A. ter Meulen (Eds.), *Handbook of logic and language* (pp. 93–177). North Holland / Elsevier.

Morrill, G. (1994). *Type logical grammar: Categorial logic of signs.* Springer.

Nivre, J., Agić, Ž., Ahrenberg, L., & et al. (2017). *Universal dependencies 2.0.* (LINDAT/CLARIN digital library at the Institute of Formal and Applied Linguistics (ÚFAL), Faculty of Mathematics and Physics, Charles University)

Nivre, J., De Marneffe, M.-C., Ginter, F., Goldberg, Y., Hajic, J., Manning, C. D., ... others (2016). Universal dependencies v1: A multilingual treebank collection. In *LREC.*

Nivre, J., Hall, J., Nilsson, J., Chanev, A., Eryigit, G., Kübler, S., ... Marsi, E. (2007). Maltparser: A language-independent system for data-driven dependency parsing. *Natural Language Engineering, 13*, 95–135.

Padó, S., & Lapata, M. (2007). Dependency-based construction of semantic space models. *Computational Linguistics, 33*(2), 161–199.

Park, H., Kang, J.-S., Choi, S., & Lee, M. (2013). Analysis of cognitive load for language processing based on brain activities. In M. Lee, A. Hirose, Z.-G. Hou, & R. M. Kil (Eds.), *Neural information processing* (pp. 561–568). Berlin, Heidelberg: Springer Berlin Heidelberg.

Pauls, A., & Klein, D. (2012). Large-scale syntactic language modeling with treelets. In *Proceedings of the 50th annual meeting of the association for computational linguistics* (pp. 959–968). Jeju, Korea.

Pennington, J., Socher, R., & Manning, C. (2014). GloVe: Global vectors for word representation. In *Proceedings of the 2014 conference on empirical methods in natural language processing (EMNLP)* (pp. 1532–1543). Doha, Qatar: Association for Computational Linguistics.

Pereira, F. (2000). Formal grammar and information theory: Together again? In *Philosophical transactions of the royal society* (p. 1239-1253). London: Royal Society.

Pollard, C., & Sag, I. A. (1994). *Head-driven phrase structure grammar.* Stanford, CA: CSLI Publications.

Pullum, G. K., & Scholz, B. C. (2001). On the distinction between model-theoretic and generative-enumerative syntactic frameworks. *Logical aspects of computational linguistics: 4th international conference, Le Croisic, France, June 27–29, 2001, Proceedings.*

Radford, A., Narasimhan, K., Salimans, T., & Sutskever, I. (2018). Improving language understanding by generative pre-training. *OpenAI.*

Richardson, K., Hu, H., Moss, L. S., & Sabharwal, A. (2020). Probing natural language inference models through semantic fragments. In *The thirty-fourth AAAI conference on artificial intelligence, AAAI 2020, the thirty-second innovative applications of artificial intelligence conference, IAAI 2020, the tenth AAAI symposium on educational advances in artificial intelligence, EAAI 2020, New York, NY, USA, February 7–12, 2020* (pp. 8713–8721). AAAI Press.

Rumelhart, D. E., McClelland, J. L., & PDP Research Group, C. (Eds.). (1986). *Parallel distributed processing: Explorations in the microstructure of cognition, Vol. 1: Foundations.* Cambridge, MA, USA: MIT Press.

Schrimpf, M., Blank, I., Tuckute, G., Kauf, C., Hosseini, E. A., Kanwisher, N., ... Fedorenko, E. (2020). Artificial neural networks accurately predict language processing in the brain. *bioRxiv.*

Shieber, S. M. (1985). Evidence against the context-freeness of natural language. *Linguistics and Philosophy, 8*(3), 333–343.

Shieber, S. M. (2004). *The turing test: Verbal behavior as the hallmark of intelligence.* The MIT Press.

Shieber, S. M. (2007). The turing test as interactive proof. *Nous, 41*(4), 686-713.

Shieber, S. M., & Schabes, Y. (1990). Synchronous tree-adjoining grammars. In *13th international conference on computational linguistics, COLING 1990, University of Helsinki, Finland, August 20–25, 1990* (pp. 253–258).

Smolensky, P. (1990). Tensor product variable binding and the representation of symbolic structures in connectionist systems. *Artificial Intelligence, 46*(1–2), 159–216.

Socher, R., Karpathy, A., Le, Q. V., Manning, C. D., & Ng, A. Y. (2014). Grounded compositional semantics for finding and describing images with sentences. *Transactions of the Association for Computational Linguistics, 2.*

Socher, R., Pennington, J., Huang, E. H., Ng, A. Y., & Manning, C. D. (2011, July). Semi-supervised recursive autoencoders for predicting sentiment distributions. In *Proceedings of the 2011 conference on empirical methods in natural language processing* (pp. 151–161). Edinburgh, Scotland, UK: Association for Computational Linguistics.

Socher, R., Perelygin, A., Wu, J., Chuang, J., Manning, C. D., Ng, A., & Potts, C. (2013, October). Recursive deep models for semantic compositionality over a sentiment treebank. In *Proceedings of the 2013 conference on empirical methods in natural language processing* (pp. 1631–1642). Seattle, WA, USA: Association for Computational Linguistics.

Solaiman, I., Brundage, M., Clark, J., Askell, A., Herbert-Voss, A., Wu, J., ... Wang, J. (2019). Release strategies and the social impacts of language models. *ArXiv*, *abs/1908.09203*.

Sprouse, J., & Almeida, D. (2013). The empirical status of data in syntax: A reply to Gibson and Fedorenko. *Language and Cognitive Processes*, *28*, 222–228.

Sprouse, J., Schutze, C., & Almeida, D. (2013). A comparison of informal and formal acceptability judgments using a random sample from Linguistic Inquiry 2001–2010. *Lingua*, *134*, 219–248.

Sprouse, J., Yankama, B., Indurkhya, S., Fong, S., & Berwick, R. C. (2018). Colorless green ideas do sleep furiously: Gradient acceptability and the nature of the grammar. *The Linguistic Review*, *35*(3), 575–599.

Steedman, M. (2000). *The syntactic process.* Cambridge, MA: MIT Press.

Sutton, P. (2017, 06). Probabilistic approaches to vagueness and semantic competency. *Erkenntnis*, *83*, 1–30.

Sweller, J. (1988). Cognitive load during problem solving: Effects on learning. *Cognitive Science*, *12*(2), 257–285.

Tai, K. S., Socher, R., & Manning, C. D. (2015, July). Improved semantic representations from tree-structured long short-term memory networks. In *Proceedings of the 53rd annual meeting of the association for computational linguistics and the 7th international joint conference on natural language processing (volume 1: Long papers)* (pp. 1556–1566). Beijing, China: Association for Computational Linguistics.

Talman, A., & Chatzikyriakidis, S. (2019, August). Testing the generalization power of neural network models across NLI benchmarks.

In *Proceedings of the 2019 ACL workshop BlackboxNLP: Analyzing and interpreting neural networks for NLP* (pp. 85–94). Florence, Italy: Association for Computational Linguistics.

Turing, A. M. (1950). Computing machinery and intelligence. *Mind, 59*(October), 433–460.

Turney, P., & Pantel, P. (2010). From frequency to meaning: Vector space models of semantics. *Journal of Artificial Intelligence Research, 37*.

Valiant, L. G. (1984). A theory of the learnable. In *Proceedings of the sixteenth annual ACM symposium on theory of computing (STOC '84)* (pp. 436–445). New York, NY, USA: ACM Press. doi: http://doi.acm.org/10.1145/800057.808710

Vaswani, A., Shazeer, N., Parmar, N., Uszkoreit, J., Jones, L., Gomez, A. N., ... Polosukhin, I. (2017). Attention is all you need. In *Proceedings of the 31st international conference on neural information processing systems* (pp. 6000–6010). Red Hook, NY, USA: Curran Associates Inc.

Vinyals, O., Toshev, A., Bengio, S., & Erhan, D. (2014). Show and tell: A neural image caption generator. *ArXiv, abs/1411.4555*.

Wang, A., Singh, A., Michael, J., Hill, F., Levy, O., & Bowman, S. (2018). GLUE: A multi-task benchmark and analysis platform for natural language understanding. In *Proceedings of the 2018 EMNLP workshop BlackboxNLP: Analyzing and interpreting neural networks for NLP* (pp. 353–355). Brussels, Belgium: Association for Computational Linguistics.

Warstadt, A., Singh, A., & Bowman, S. (2019). Neural network acceptability judgments. *Transactions of the Association for Computational Linguistics, 7*(0).

Williams, A., Drozdov, A., & Bowman, S. R. (2018). Do latent tree learning models identify meaningful structure in sentences? *Transactions of the Association for Computational Linguistics, 6*, 253–267.

Williams, A., Nangia, N., & Bowman, S. (2018). A broad-coverage challenge corpus for sentence understanding through inference. In *Proceedings of the 2018 conference of the north American chapter of the association for computational linguistics: Human language technologies, Volume 1 (long papers)* (pp. 1112–1122). New Orleans, LA: Association for Computational Linguistics.

Xu, K., Ba, J., Kiros, R., Cho, K., Courville, A., Salakhudinov, R., ... Bengio, Y. (2015). Show, attend and tell: Neural image caption generation with visual attention. In F. Bach & D. Blei (Eds.), *Proceedings of machine learning research* (Vol. 37, pp. 2048–2057). Lille, France: PMLR.

Yang, Z., Dai, Z., Yang, Y., Carbonell, J. G., Salakhutdinov, R., & Le, Q. V. (2019). XLNet: Generalized autoregressive pretraining for language understanding. *ArXiv, abs/1906.08237*.

Yogatama, D., Blunsom, P., Dyer, C., Grefenstette, E., & Ling, W. (2017). Learning to compose words into sentences with reinforcement learning. In *5th international conference on learning representations, ICLR 2017, Toulon, France, April 24–26, 2017, Conference track proceedings*.

Zhang, W. E., Sheng, Q. Z., Alhazmi, A., & Li, C. (2020). Adversarial attacks on deep-learning models in natural language processing: A survey. *ACM Transactions on Intelligent Systems and Technology, 11*(3).

Author Index

Subject Index

Printed in the United States
by Baker & Taylor Publisher Services